The American Society of Agronomy: 100 Years of History

Editor:
Lowell E. Moser

Prophetic Voices from Our Past

Creators/Designers:
Vivien Gore Allen and Philip Brown

ASA Centennial Committee:
Lowell E. Moser, Chair
Vivien Gore Allen
Douglas Karlen
Martin Massengale
Murray Milford
Calvin Pearson
Arthur Peterson
Dwayne Rohweder
Sara Uttech, Ex Officio
Lisa Al-Amoodi, Ex Officio

**American Society of Agronomy, Inc.
Madison, Wisconsin, USA**

2007

Cover: An early use of contour farming coupled with a contemporary aerial view of contour buffer strips in northeast Iowa. USDA-NRCS.

Copyright © 2007 by the American Society of Agronomy, Inc.

ALL RIGHTS RESERVED. No part of this publication may be reproduced or transmitted in any form or by any means, electronic or mechanical, including photocopying, recording, or any information storage and retrieval system, without permission in writing from the publisher.

The views expressed in this publication represent those of the individual Editors and Authors. These views do not necessarily reflect endorsement by the Publisher(s). In addition, trade names are sometimes mentioned in this publication. No endorsement of these products by the Publisher is intended, nor is any criticism implied of similar products not mentioned.

Additional photo credits: USDA-NRCS, USDA-ARS, NASA, University of Wisconsin-Madison Department of Agronomy, Curtis Publishing, Imageafter.com.

American Society of Agronomy, Inc.
677 South Segoe Road, Madison, WI 53711-1086 USA

ISBN: 978-0-89118-166-8

Printed on recycled paper in the United States of America.

CD: The CD is designed for Windows and PopwerPoint and will autorun for PC users. Mac users can access the Voices presentation by going to: \AutoPlay\Docs\ASA100Voices2003.pps and can view the PDF files and WAV files in the same Docs folder.

Keep history alive. Please donate this book to your library when you are finished with it.

Contents

Foreword — iv
Jerry L. Hatfield

Preface — v
Lowell E. Moser

OVERVIEW

The American Society of Agronomy: An Overview of One Hundred Years — 3
Lowell E. Moser

MARK ALFRED CARLETON

Mark Alfred Carleton, President, American Society of Agronomy 1907–1908 — 13

The Agronomic Legacy of Mark A. Carleton — 15
Gary M. Paulsen

HISTORICAL ACCOUNTS

History of the Organization of the American Society of Agronomy — 21
T. Lyttleton Lyon

History of American Society of Agronomy First Fifty Years—1907 to 1957 — 24
H.H. Laude

Development of the American Society of Agronomy, 1958–1977 — 35
D.C. Smith

Building the ASA Headquarters in Madison — 45
Dale Smith

Development of the American Society of Agronomy, 1978–2007 — 47
Dwayne A. Rohweder, Robert F Barnes, David M. Kral, and Calvin O. Qualset

Agronomy Journal **Turns One Hundred** — 61
Calvin H. Pearson, Susan M. Ernst, Ken A. Barbarick, Jerry L. Hatfield, Gary A. Peterson, and Dwayne R. Buxton

RESIDENT EDUCATION

Development and Evolution of Resident Education in the American Society of Agronomy — 71
M.H. Milford and D.T. Smith

EXTENSION EDUCATION

The American Society of Agronomy and Development of Agronomic Extension in the United States — 79
Dwayne A. Rohweder

ASA IN THE WORLD

The International Dimension of the American Society of Agronomy: Historical Perspective, Issues, and Challenges — 89
W.A. Payne and J. Ryan

Foreword

Celebration of the achievements of the American Society of Agronomy over a 100-year period represents the contributions of many individuals. The American Society of Agronomy was formed as the vision of Mark Carleton; however, the contributions of many people since that time have made ASA what it is today and what it will be tomorrow. There have been many individuals who have helped shape ASA through their interests in education, extension, research, policy, and publications. It is important to review our history, which is what is contained in this overview of the Society, but to also view this effort as the foundation for how we move forward with dedication and determination to continue to fulfill the vision of sharing agronomy with the world.

We are indebted to the efforts of Lowell Moser and the Centennial Committee to bring this volume together and compile the efforts of many of our great and valuable ASA members.

It is my pleasure to present this history of the American Society of Agronomy to you, and to serve you as ASA's 100th president.

Jerry L. Hatfield
President, American Society of Agronomy

Preface

In planning for the publication celebrating the 100-year history of the American Society of Agronomy, we decided to prepare a publication highlighting different periods and aspects of ASA. In this publication we have assembled much of the historical information that is available for ASA. We have brought together the histories that have been written by those who lived the history, the earlier ones by those who preceded us, and the more current ones by those who are most familiar with it and lived it.

In the first chapter, an attempt has been made to present a brief overview that provides a broad historical picture of the 100 years without getting into much detail. The next section describes our first president, Mark A. Carleton, and his contributions. Through his vision and leadership ASA was formed. His background was first published in the *Proceedings of the American Society of Agronomy* in 1911, and his agronomic legacy is illustrated in an article by Paulsen published in 2001 in the *Journal of Natural Resources and Life Sciences*. The next section compiles historical accounts, beginning with a piece written by T. Lyttleton Lyon, charter member, first secretary, 11th president, and 1933 historian of ASA. It describes the founding of ASA and history of the first 25 years. Next is a comprehensive history of the first 50 years (1907–1957) developed by H.H. Laude, with sections written by authors active in the various areas, and published in *Agronomy Journal* in 1962. Also included is a report of the development of the headquarters and building in the 1960s, written by Dale Smith, a member of the ASA Building Committee. A detailed description of the development of ASA from 1958 until 1977, written by D.C. Smith, follows. To complete the century of ASA history, Dwayne Rohweder (member for 54 years), Robert Barnes (member for 53 years and former ASA–CSSA–SSSA Executive Vice President), David Kral (former Associate Executive Vice President), and Cal Qualset (ASF Board Chair) summarize ASA's development for the last 30-year period (1978–2007).

The articles from Society journals were reproduced largely intact. A small number of inconsistencies in style of presentation were made uniform, but the articles are presented as close as possible in their original form. Some material may have been added, such as photos or notes, and in some instances the articles are "reprinted in part." In these cases, the reader is directed to the original article for the complete content.

The rest of the chapters were written specifically for this 100-year historical summary and are organized by topic. The history of *Agronomy Journal* is provided by Calvin Pearson, the current editor. Murray Milford and Dudley Smith provide a history of resident education and student activities of ASA. Agronomic Extension and the role of ASA are summarized by Dwayne Rohweder. Bill Payne and John Ryan discuss the international aspect and development of ASA.

The Prophetic Voices from Our Past multimedia program, created by Vivien Gore Allen and Philip Brown, features the past ASA presidents and quotes from their presidential addresses or other sources, as well as transcripts of their complete presidential addresses. "A Century of Women in Agronomy," written by Marla McIntosh and Steve Simmons, describes the entry of women into the agronomic profession and features the individual life stories of women from a number of decades in the century. In addition, the CD includes century-long lists that record Society history—award winners, annual meeting locations and dates, and other tabular data too long for inclusion in the book. Our goal in bringing these two historical publications together in one book and CD package is to provide members and others first-hand accounts of the history and to preserve important historical knowledge and documents to prevent their being lost or buried because of time or location. We believe this book and CD will be a valuable historical reference of the centennial history of the American Society of Agronomy.

Lowell E. Moser
Chair, ASA Centennial Committee

Overview

The American Society of Agronomy: An Overview of One Hundred Years

Late in the nineteenth and early in the twentieth century departments of agriculture in agricultural colleges were divided into more specific units (Lyon, 1933, also reprinted in part in this publication, p. 21–23). One of these units was Agronomy or Crops and Soils. Very early in the twentieth century the United States Department of Agriculture (USDA) Bureau of Plant Industry (BPI) began to use the term *agronomist* in position titles. Whereas there were only 16 agronomists at Land Grant Colleges and the title was not used by the USDA in 1900, there were 59 state-supported agronomists and more 100 agronomists in the USDA by 1908 (Carleton, 1910). During 1907, an agronomic seminar met at the BPI in Washington, DC. At one of these seminars in October, 1907, a discussion ensued about the formation of a national society (Lyon, 1933). A committee consisting of M.A. Carleton, W.J. Spillman, C.V. Piper, E.C. Chilcott, and A.D. Shamel was formed and instructed to obtain the opinions of agronomists and administrators, including college presidents, regarding the feasibility of forming a national agronomy organization. On November 30, 1907, this committee sent a letter (Lyon, 1933) to U.S. agronomists and administrators inviting them to join a final committee, which would call a meeting at the American Association for the Advancement of Science (AAAS) in Chicago in late 1907, to explore the formation of such an organization. The committee issuing this call indicated that they would have everything ready, including a draft of a constitution and bylaws, for the formation of this agronomic society. There were 58 replies. Some questioned the advisability of adding another dues-paying organization, but the agronomists unanimously agreed that a society of agronomists should be formed (Lyon, 1933). After the positive feedback a second letter was sent on December 12, 1907 (ASA, 1910, p. 6) containing the names of 38 people willing to attend the meeting in Chicago and assist in forming the society.

On December 31, 1907, 43 people (Miller, 1962) met in the Botany Building at the University of Chicago. Mark A. Carleton was appointed chairman and T. Lyttleton Lyon was appointed secretary (ASA, 1910, p. 7). The meeting was called to order at 9:00 a.m. Extensive and varied discussion followed. A motion was made that a Committee on Permanent Organization be formed, with M.A. Carleton as chair. Messrs. Armsby, Hopkins, Piper, and Lyon comprised the committee. The meeting was recessed until 2:00 p.m. At that time the committee presented a proposal to form a society, including a preliminary constitution and bylaws. The proposed name of the organization was "The American Society of Agronomy," and its objective would be "the increase and dissemination of knowledge concerning crops and soils and the conditions affecting them." Dues were set at $2.00 annually. It was agreed that all persons joining before July 1, 1908, would be considered charter members. The first list of members with the charter members indicated was published in Proceedings of American Society of Agronomy, Volume 2. (ASA, 1911, p. 19–23). Also, local sections consisting of 10 or more members could be developed. Those present adopted the committee report and the American Society of Agronomy came into existence. The new Society elected its first set of officers as

Lowell E. Moser
Department of Agronomy and Horticulture
University of Nebraska, Lincoln

Left: An Iowa farm. Farmers through time have looked out over their fields and pondered their progress and what the future may bring, a fitting metaphor as we look at the history and future of ASA. Photo by USDA-NRCS.

Above: The seal of the American Society of Agronomy.

Section page: Breaking sod in the early 1900s. Photo courtesy of the University of Wisconsin Agronomy Department.

Corresponding author
L. Moser, Department of Agronomy and Horticulture, University of Nebraska
P.O. Box 830915, Lincoln, NE 68483-0915 (email: lmoser@unlnotes.unl.edu).

follows: M.A. Carleton, President; C.P. Bull, First Vice President; J.F. Duggar, Second Vice President; T.L. Lyon, Secretary; E.G. Montgomery, Treasurer. On the following day, January 1, 1908, the first scientific meeting of the society was held. Ten papers were presented, four by title only (Lyon, 1933). A regional meeting was held at Cornell University, Ithaca, NY in July 1908. The first national annual meeting following the organizational meeting was held in Washington, DC November 17–18, 1908. Thus began the hundred-year history of annual meetings of the ASA (a complete list of the meetings and locations is available on the accompanying CD).

PUBLICATIONS
Agronomy Journal

Mention of a publication was made at the Ithaca, NY meeting in 1908, and a publication committee was appointed at the December 7–8, 1909, meeting in Omaha, NE with the authorization to proceed with the printing of the proceedings of papers given at the first four meetings, 1907–1909. Members were C.V. Piper (chair), C.R. Ball (secretary), G.N. Coffey, G.H. Failyer, and L.H. Smith. Mr. C.R. Ball collected and edited the papers that had been presented at the first four meetings (January 1908, July 1908, November 1908, and December, 1909) for Volume 1 of the *Proceedings of the American Society of Agronomy* published in 1910. Volume 2 (1910) was published in 1911. The Proceedings continued for two more years, 1911 and 1912, and contained a few contributed papers not presented at the annual meetings (Lyon, 1933).

Recognizing the need for a regularly published journal, the publication of the *Journal of the American Society of Agronomy* (JASA) was authorized in 1912, and on August 8, 1913, the first issue of JASA (Vol. 5) was published. At the beginning, the JASA was envisioned as a quarterly publication, and the sum of $1000 annually was thought to be adequate to publish the journal. To meet publication demand JASA went to bimonthly publication. From 1915 to 1921 (C.W. Warburton's term as editor) it went to seven, eight, and finally nine issues per year (Luckett, 1962). In 1923, the JASA began monthly publication and continued for 38 years until 1961. From 1961 until 2007, six issues have been published per year. Starting with Volume 41 in 1949 the name of the JASA was changed to *Agronomy Journal* (AJ), and the 5 by 7 inch format was changed to an 8 ½ by 11 inch format. A special "Golden Jubilee" issue of the AJ (Vol. 29, Issue 12) was published in December 1957 and contains historical articles and Golden Jubilee meeting information. For details on *Agronomy Journal*, including editors, subscriptions, and challenges, see Lyon (1933, Luckett (1962), Fuccillo (1983), and Pearson et al. (2007, this publication). From the very beginning publishing a highly respected journal has always been a top priority of ASA. In 1998 AJ began electronic publication; online hosting by HighWire Press began in 2001. Continuous online publication began in 2005. In 2008, the entire archive of back issues of AJ (as well as *Crop Science, Journal of Environmental Quality*, and *Soil Science Society of America Journal*) are being made available online through HighWire through their existing sites at www.scijournals.org.

Soil Science Society of America Journal

At the 1936 annual meeting when the Soil Science Society of America was formed, the *Proceedings of the Soil Science Society of America* was initiated. It began as an annual proceedings where papers from the annual meetings were published "as soon as possible after the annual meeting." Emil Truog, C.E. Millar, F.J. Alway, and Richard Bradfield (chair) comprised the initial editorial committee. In 1952, (Vol. 16) publication was changed to quarterly. (With this transition the year 1951 was omitted from the series.) In 1957 publication increased to six issues annually. Beginning with Vol. 40 in 1976, the name was changed to the *Soil Science Society of America Journal*, and it has continued as a bimonthly publication until the present, with continuous online posting beginning in 2005.

Crop Science

The Crop Science Society of America was formed in 1955, and the first issue of the new journal *Crop Science* was published in 1961. Dr. I.J. Johnson was the first editor. It has continued as a bimonthly publication. At the same time (1961) *Agronomy Journal* changed from monthly to bimonthly publication. Continuous online posting began in 2005.

Journal of Agronomic Education

A demand existed for an outlet for articles addressing teaching and education for both resident and extension education. There were generally more articles of this nature than the *Agronomy Journal* had room to publish, and with an editorial board dominated by research scientists, many educational articles were not accepted. The Board of Directors voted in 1970 to establish a new journal devoted to agronomic education. The first issue of the *Journal of Agronomic Education* (JAE), sponsored by ASA, appeared in October, 1972. Drs. W.F. Hueg and B.R. Bertramson were the first editors. Educational research papers, symposia, addresses, and other papers that would not be accepted by *Agronomy Journal* were welcome in JAE. From 1972 until 1983, JAE was published annually. In 1984 (Vol. 13) the JAE was published twice a year, spring and fall. Two issues were published annually until 1998, when it dropped back to one issue per year. In 1992, the title was changed to the *Journal of Natural Resources and Life Science Education* (JNRLSE) to attract a broader range of papers and educators. In 1998, the JNRLSE began electronic publication, posting articles online regularly, with a hard copy and CD compilation at the end of the year.

Journal of Environmental Quality

In 1972, the Societies sponsored the publication of the *Journal of Environmental Quality* (JEQ). The objectives of this new journal were to (i) provide focus on environmental quality work which the leadership viewed is lacking in the other journals at the time, (ii) provide recognition for scientists working in the environmental quality area, and (iii) make it possible for other scientists in other disciplines to locate and recognize our contributions in environmental quality (Black et al., 1972). It was published quarterly until 1994, when it went to a bimonthly publication schedule. In 2002, HireWire Press began electronic publication of JEQ, and continuous publication began in mid 2005.

Examples of Agronomy Journal *through the years.*

Journal of Production Agriculture

The *Journal of Production Agriculture* began publication in January 1988, as an ASA–CSSA–SSSA publication with five other cooperating organizations. It was a technical peer-reviewed journal and was envisioned to "fill a demand for information of more immediate, applicable level and would provide an opportunity for interaction with other agricultural disciplines" and to help "transmit the information in an understandable and useful manner (Gast et al., 1988). Robert Hoeft was the first editor. To help "transmit the information in an understandable and useful manner" Research Application Summaries were added in 1993 (Vol. 6). In this special section of the journal, each article was abstracted and related to the potential application by posing practical questions and answering them with research results. However, circulation of this journal was never very high, and it operated at a loss. Publication ceased in 1999 (Vol. 12).

Plant Management Network Journals

In 2002, CSSA and ASA initiated *Crop Management*, an online journal designed for practicing agronomists, through the Plant Management Network (PMN, www.plantmanagementnetwork.org) founded by the American Phytopathological Society (APS). A second PMN journal, *Forage and Grazing Lands*, was started in 2003 with ASA and CSSA sponsorship, and a third PMN electronic journal, *Applied Turfgrass Science* was initiated in 2004, also with ASA and CSSA sponsorship.

Crops and Soils

In 1945, one of the Society's objectives was to develop a publication containing articles on the practical application of agronomy which would be of interest to extension personnel, high school Vo-Ag teachers, producers, industrial agronomists, teachers, and others who were interested in short, practical articles. The resulting publication was *What's New in Crops and Soils*, later simply called *Crops and Soils*. The launching of this publication coincided with the formation of a permanent headquarters. Mr. L.G. Monthey was employed by the ASA to be the editor for *What's New in Crops and Soils* and the executive secretary of the two Societies (ASA and SSSA). The first issue of *Crops and Soils* was published in October 1948. Nine issues were published each year. It was published monthly from October through March and bimonthly from April through September. The magazine initially had 8500 subscribers, a number that grew to more than 25,000 in 1972–1973. Although the magazine enjoyed relatively good popularity, it often lost money for the Societies and subscribers were down to 18,300 in 1987. It was discontinued with the seventh issue in 1987 (Vol. 39) because of large economic losses. William Luellen (1987), editor of *Crops and Soils*, wrote a fond "adieu" to the subscribers of *Crops and Soils* magazine and indicated that a new applied agricultural publication, *Journal of Production Agriculture* would fill the void.

With increasing numbers of crop advisers and others interested in practical agronomy, interest in semipopular publication persisted into the 21st century. In 2006, the ASA Board approved the reviving of *Crops and Soils* as a quarterly publication, and publication began with the spring issue in March 2007 (Vol. 40). The new *Crops and Soils* (https://www.agronomy.org/cropsandsoils/) contains practical articles, industry and member news, and ways for Certified Crop Advisers to obtain continuing education units by taking quizzes about selected articles.

CSA News

Originally "Agronomy Affairs" or "Agronomy News" was published as an integral part of most issues of *Agronomy Journal* to provide news of interest to members. In 1956, *Agronomy News* became a free-standing, bimonthly publication, serving all three Societies. Bimonthly publication continued until October 1983, when monthly publication began. The title was changed in 1998 to *Crop Science, Soil Science, Agronomy News*. In 2000 the title was shortened to *CSA News*. In 2006, *CSA News* was changed to a slick-paper, color magazine. Over the years it continued to carry agronomic and member news, columns from officers and Society staff and position announcements. Now, in addition, it contains subject matter articles as well as featuring upcoming and "most-read" articles from the journals. This publication has always been well received, and members look forward to the new issue each month.

Books

Agronomy Monographs. Although a monograph committee was appointed in 1941, the first monograph, *The Colloid Chemistry of Silicate Materials,* prepared by C.E. Marshall, was published in 1948. The first six monographs were published by Academic Press; thereafter ASA published its own monographs. In 1974, CSSA and SSSA began to cosponsor monographs. As of 2007, there have been 49 monograph titles published summarizing the state of knowledge of a wide range of subjects (refer to the list of publications on the accompanying CD). A number have been revised, and second and third editions of some monographs exist.

Special Publication Series. The ASA Special Publication Series was initiated in 1962. Special publications were initiated to meet publication needs of the Society that were not met through the regular publications (Smith, 1980). These publications generally assembled the papers from symposia or other meetings. There were 66 special publications sponsored by ASA, CSSA, and/or SSSA, as of 2003.

Other Books. In 1975, a textbook series, Foundations for Modern Crop Science, was begun with *Crops and Man*, edited by J.R. Harlan. The goal of these books was to organize subject matter and relate it to current discoveries and new principles to stimulate student interest. The ASA also has collaboratively published with other groups numerous other books summarizing special research meetings (see a list of publications on the accompanying CD). As of 2007, the Societies, through a combined ASA–CSSA–SSSA Book and Multimedia Publishing Committee, are working to expand the scope of the publishing program. Their work includes exploring new modes of publishing, from short-run printings to digital offerings, as well as expanding the scope of disciplines covered and the target audiences, such as students and the general public.

ORGANIZATION
ASA and the Development of SSSA and CSSA

At the outset the American Society of Agronomy was comprised of individuals interested in crops, soils, or both. At the first meeting on January 1, 1908, two individuals were appointed as a program committee, C.G. Williams for Crops and A.R. Whitson for Soils. By 1924, annual meeting program sessions were categorized as Soils or Farm Crops. In 1930, a committee was appointed to look into the organization of the ASA. The recommendations in their 1931 report (ASA, 1931, p. 1032–1034) were to retain the American Society of Agronomy name and to divide into two sections, Crops and Soils. Each section was to work out subsections, and each section would elect a chair and a vice-chair. The 1936 annual meeting for Soils was jointly planned by the Soils Section of ASA and the American Soil Survey Association. In the Soils Section there were six subsections listed for 1936: I. Soil Physics, II. Soil Chemistry, III. Soil Microbiology, IV. Soil Fertility, V. Soil Genesis, and VI. Soil Science Applied to Land Use. At the 1936 annual meeting, the Soils Section and the American Soil Survey Association (ASSA) voted to merge. This action made the six sections in the Soils Division in ASA and the ASSA the Soil Science Society of America (SSSA). The name of the sixth section changed to Soil Technology. Some divisions have changed names and new divisions have been added. In 2007 there were 11 divisions in SSSA.

Although the Crops Section in ASA was formed in 1931, no subsections were developed until 1937, with the formation of these sections: I. Genetics, Cytology, and Breeding; II. Physiology, Morphology, and Ecology; and III. Miscellaneous Topics. In the mid 1940s Subsection III became the Agronomic Application subsection. In 1946 Divisions began to be used for major groups in the Society and subdivisions for the next level of groupings, although this use was not always consistent. The Crops Division was reorganized into five subdivisions: I. Breeding, Genetics, and Cytology; II. Physiology and Ecology; III. Production and Management; IV. Seed Production and Technology; and V. Special Topics. In 1947 Subdivision V was divided into V. Turf and VI. Weed Control, and Special Topics was dropped. In 1949, a committee was formed to explore the possibility of forming a Crop Science Society of America (CSSA) from the Crops Division of ASA. In 1953 bylaws were developed. They were approved at the 1954 annual meeting, forming the CSSA for 1955. The divisions remained the same except the sixth section name was changed to Weeds and Weed Control. Divisions in CSSA have been added and modified. In 2007 there are eight divisions in CSSA. However, the names of the three Societies have remained the same since their inceptions, 100 (ASA), 71 (SSSA), and 52 (CSSA) years ago.

Divisional Structure

A major reorganization of ASA was approved by members in 1947 (Keim, 1953). In addition to the reorganization of the Crops Division, Agronomic Education was formed as a provisional division (XIII). It became permanent in 1949 with three sections: 1. Resident Teaching, 2. Extension Participation, and 3. Student Activities. A temporary division of Agronomic Applications with four sections (Pastures, Soil Conservation, Turf, and Seed Production) and a temporary division of Plant Nutrients were also added. The ASA Coordinating Committee made an extensive report at the 1950 annual meeting (ASA, 1951, p. 52–55) on restructuring and renaming the Society. They proposed six models of organization and six names for the overall Society, including the current one. This was sent to the membership for vote by a mail ballot. The results (ASA, 1950, p. 630–631) showed that the membership favored retention of both the American Society of Agronomy name and the current structure (ASA, SSSA, and the Crops Division). However, the CSSA formed four years later. Although after formation of the SSSA a person could join either Society or both, later membership in ASA became automatic if a person joined either SSSA or CSSA. The three Societies continued to function with that structure for the next 50 years. Discussion began in 1997 regarding establishing clearer identities for each Society and removing SSSA and CSSA from under the ASA umbrella. Again, six models were proposed and the ASA Board at the 1998 meetings chose the model that retained the existing structure but eliminated the automatic ASA membership for CSSA and SSSA members (McFee, 1999).

In 2003, discussions began again about restructuring the Societies and reducing the size of the boards. The SSSA and CSSA adopted new bylaws reducing the size of their board in 2005. Once that was completed ASA changed their bylaws in 2005 to reduce their board from 46 to 16. The C and S divisions and regional branch representatives no longer are members of the ASA Board. The result of these actions is the complete independence of each of the Societies so they are parallel societies, each with a board of directors that can concentrate on respective Society programs. Since formation of the separate Boards of Directors, the Societies are in the process of establishing a management unit that will address the management of the three Societies and any others that may affiliate in the future. This unit, the Alliance of Crop, Soil, and Environmental Science Societies (ACSESS), will hold the assets, employ the staff, and handle the business affairs of the associated Societies. As of the centennial meeting formation of ACSESS has not been completed.

Establishment of Additional ASA Divisions

The American Society of Agronomy contained all of the agronomy, crops, and soils divisions since they were first established in the 1930s until the reorganization action in 2005. Action by the Soils Division in 1936 designated the first six divisions as the Soil Science Society of America and, likewise, action by the Crops Division in 1954 designated divisions seven through twelve as the Crop Science Society of America. The continuous numbering system for both the crops, soils, agronomy sections (I–XIII) was used until in the 1963 meeting program ASA, CSSA, and SSSA subject matter divisions were designated as A, C, and S, respectively, and were numbered separately.

Division XIV Land Use and Management and the subdivison Military Lands were formed in 1956 and in 1963 merged and became Division A-2, Land Use and Management. The name Miliary Land Use and Management was adopted in 1984 for A-2. Division A-3, Meteorology and Climatology was formed in 1963. It was renamed Agroclimatology and Agronomic Modeling in 1978. In 1947 the Agronomic Education Division developed three subdivisions, Resident Teaching, Extension Participation, and Student Activities. In 1964 the division formed two provisional divisions. A-1 was entitled Resident Education Teaching and had one subdivision, Student Activities. Division A-4 was entitled Extension Education. Both divisions became permanent in 1967. Division A-5, Environmental Quality, was added in 1971 and made permanent in 1974. Division A-6, International Agronomy, was added provisionally in 1972 and became permanent in 1975. Division A-7, Agricultural Research Station Management, was added in 1981 and became permanent in 1984. Division A-8, Plant Science Applications, was formed provisionally in 1990. The name was changed to Integrated Agricultural Systems in 1995, and the division became permanent. Division A-9, Professional Practitioners (Steering Committee), was established provisionally in 2000, and in 2002 Professional Practitioners became the name of the permanent A-9 Division.

Branches

The original ASA bylaws provided for the formation of local sections (but not regional branches), but by 1912, five regions already were suggested. Many local sections were established by 1918, often at Land Grant colleges. During WW I most were discontinued, but some evolved into the official branches of ASA. Originally, the Western Agronomic Workers, organized in 1916, became the Western Branch of the ASA in 1924. The Western Society of Soil Science was formed in 1921. The Western Branch of Soil Science was not affiliated with ASA but officially

The leadership has shaped the Society over the years to ensure its structure meets the demands of its dynamic membership profile. A 2003 board meeting, with (from left) Past President Charlie Stuber, the author and President-Elect Lowell Moser, President Bob Hoeft, and EVP Ellen Bergfeld.

affiliated with SSSA in 1954. The Western Agronomic Workers became the Western Branch of ASA and changed their name to the Western Society of Crop Science in 1954. The New England local section was first formed in 1915, became inactive during WWI, and was reactivated in 1921. In 1930, it reorganized and became an official branch of ASA; the name was changed to the Northeastern Section in 1931. The Agronomy Section of the Association of Southern Agricultural Workers, formed in 1921, first affiliated with ASA in 1926. In 1923, the Corn Belt Section was officially approved. In 1954, the ASA Constitution was revised and the regional sections became the regional branches. In 1954 the Corn Belt Section changed its name to the North Central Branch. As of 2007, the Northeastern, North Central, Southern, and Western branches are official affiliates of ASA. The Western branch is comprised of the Western Society of Crop Science and Western Society of Soil Science, and they meet together occasionally. In 1978, CSSA recognized the Western Society of Crop Science as a CSSA branch. The SSSA reconfirmed the Western Society of Soil Science as a branch of SSSA in 2003. The Northeastern branch also was approved as a CSSA branch in 2005, becoming an affiliate of all three Societies.

A busy career placement service, 1967 Annual Meeting, Washington, DC. Networking for information sharing and enhancing the profession has always been a priority in the Society.

The branches provide regional programming, but the main focus of most members has been with the national organization, its services, and annual meetings. The branches are only loosely affiliated with the national organization. They plan their own programs and have developed their own bylaws. The only requirement is that their bylaws not conflict with those of ASA.

HEADQUARTERS

In 1946, the Policy and Program Committee made eight recommendations that had a major shaping effect on the Societies (ASA, 1946, p. 1126–1129). One was to begin a popular agronomic magazine and another was to develop a permanent headquarters. These two recommendations together resulted in hiring Lawrence G. Monthey as the editor of *Crops and Soils* and the first executive secretary of the two Societies (ASA and SSSA). A moderate-sized midwestern city was sought, and Madison, Wisconsin was selected as the location for the headquarters. In 1948, space was rented at 1910 Monroe Street. That space soon proved to be inadequate and in 1949, a suitable building was located at 2702 Monroe Street, and the Society headquarters were located there for the next 12 years, with L.G. Monthey as executive secretary. In 1958, a building committee was appointed to determine the feasibility of building a permanent headquarters for the rapidly growing Societies. In a mail ballot Madison, WI was chosen as the site over Columbus, OH or St. Paul, MN. The committee determined that for $100,000 a building could be constructed while still maintaining a comfortable cash reserve. A triangular lot was purchased between South Segoe and Odana Roads. The building contract was signed on October 25, 1961. Staff moved into the 5840-sq. ft. building on September 1, 1962, and the building was dedicated on October 29, 1962 (see Smith, 2007, this publication). Matthias Stelly became the new executive secretary/treasurer on April 1, 1961, so he was present for the final planning and construction of the building. In 1966 the Society approved the construction of a 3276-sq. ft. expansion to the building, which was completed in 1967. With the Society growing rapidly and staff activity increasing, the space limitations again became limiting in 1974. In 1975, a 6200-sq. ft. addition was authorized; the addition was completed October 13, 1976. In 1984 the building was remodeled again, including a new entrance and reception area.

ANNUAL MEETINGS

The first scientific meeting was held on January 1, 1908, one day after the formation of the Society. Although another meeting was held in July 1908 in Ithaca, NY, it was regional in nature. The second annual meeting was held November 17–18, 1908 in Washington, DC. Annual meetings of ASA have been held generally in late fall of every year except for 1944 when the meetings were cancelled because of WW II travel restrictions. The 1945 meeting also was cancelled but with the cessation of the war and lifting of travel restrictions, the 1945 annual meeting was reinstated and held February 26 to March 1, 1946 in Columbus, OH.

At early meetings, papers were more like literature reviews, and the Society had symposium-type meetings. With time and an increasing amount of crops and soils research to report, meetings became a venue for reporting original research. Symposia were still an important part of annual meeting programs. In the 1920s one of the arguments against forming subsections within the Soils and Crops Divisions was that specialization in subject matter areas could be achieved at any meeting through development of appropriate symposia (Lyon, 1933). In the early decades of the Society an important part the annual meetings was the reporting by special subject-matter committees that addressed timely problems and topics. Committees worked throughout the year, met annually, and brought together the current state of knowledge to address specific problems. Some of these committees evolved into future sections and divisions. Subject-matter committees persisted throughout most of the history of the Societies, but during the last 50 years the number of Society-wide subject matter committees has dropped, and annual meeting activity has focused on reporting research or assimilating research in symposia.

Beginning in 1962, themes were established for the annual meetings, and in 2005, each of the three Societies could develop Society-specific themes (see the list of meetings and their themes on the accompanying CD). The themes were meant to focus on broad contemporary areas and symposia, but did not limit the kinds of programs developed. Industrial exhibits were added to the annual meetings in 1931, and other activities, such as job placement, tours, student contests, and numerous others, were added to encourage meeting participation for a variety of reasons. Recently attendance has averaged about 3500, sometimes reaching 4000, and has remained high even with some decline in membership. Annual meetings are a top priority for the Societies.

AWARDS

The Society has created and presented many awards over the years, including the initiation of the Fellows program in 1925. Nearly all of them are presented at the annual meetings. For brevity the history of each award is not described here. A complete list of award recipients can be viewed on the accompanying CD.

UNDERGRADUATE STUDENT PARTICIPATION

In 1923, an ASA committee appointed to explore a national organization involving college students concluded that undergraduate student involvement would be desirable, but the formation of a student chapter should be student-initiated. In the same year a non-ASA affiliated crops judging contest was held in Chicago (Milford and Smith, 2007, this publication). However, it was not until 1932 that a student committee, comprised of representatives from midwestern clubs met at the Cornbelt section meeting to initiate the formation of a national student agronomic student organization (Milford and Smith, 2007). The national student group first met at the Chicago Crop Judging Contest in 1935 and con-

A student accepts an Agronomic Achievement Award honoring the collective efforts of his club. Presenting is Darrell Metcalf, then chair of the Student Activities Division.

tinued to meet there until 1950, when they began to meet with the ASA at their annual meetings. The ASA first sponsored the essay contest in 1933 (Metcalfe, 1962).

In the 1930s formation of ASA-affiliated student sections occurred quickly, and 23 had formed by 1941. During WW II student sections became defunct and student participation with the ASA ceased. In 1947, the Society appointed temporary officers to reinitiate the student section. Agronomy clubs at universities reactivated; by 1950 there were 30 affiliated clubs with 1287 members (Milford and Smith, 2007). The ASA student organization formally organized as a subsection, XIIIc Student Activities, under the Agronomic Education (XIII) section. After the Resident Education became a separate division, student activities became the only subdivision (1a) of the Resident Education Division. Student activity and contests continued under subdivision status for more than 50 years. After the students began meeting with ASA, contests were added and there was generally large participation by students. In 2004, students reorganized under all three Societies as the Students of Soil, Agronomic, and Environmental Sciences (SASES). Currently all three Societies have formal contact with the students, and student reports are made to each of the Society boards.

The 1961 Awards banquet.

CERTIFICATION PROGRAMS

After about 8 years of planning and preparation, the American Registry of Professionals in Agronomy, Crops, and Soils (ARCPACS) was approved in 1974 and began operation with the appointment of Martin Openshaw as director in 1977. The purpose of ARCPACS was to develop standards and procedures for the certification of qualified professionals. The areas of specialization developed were Agronomist, Crop Scientist, Crop Specialist, Soil Scientist, Soil Specialist, and Professional in Training (Smith, 1980). Specializations as a certified Plant Pathologist and Weed Scientist were added in 1991. In 1993 "ARCPACS" became the name and no longer was the acronym for the original name. Numbers of certified professionals in ARCPACS increased to about 2700 members (Rohweder et al., 2007, this publication). In 2003, the program dropped back to a single specialization, Certified Professional in Agronomy (CPAg), and the Certified Professional Soil Scientist and Classifier certifications were transferred to SSSA. In 2007, there were 658 Certified Professional Agronomists.

The state-based Certified Crop Adviser (CCA) program started in 1993, with a board for each state and a national board. The number of CCAs increased rapidly. A Canadian CCA group was developed in 1997, and the name was changed to the International Certified Crop Adviser Program (ICCA). By 2007, there were 13,568 Certified Crop Advisers.

SCIENCE POLICY

The Washington, DC program of the Societies began in 1986 when a Congressional Fellow program was initiated with the appointment of Jonathan Haskett (see the list of Congressional Science Fellows on the accompanying CD). In 1987 Dr. Terry Nipp was supported as a Congressional Fellow by the Societies and worked for Senator George Brown. A Washington, DC liaison presence was established in 1995 under a contractual arrangement with AESOP Enterprises, an independent firm headed by Terry Nipp. In 1999, the Societies established their own Science Policy Office, with Dr. Karl M. Glasener as the Director of Science Policy. A science policy intern program was started by the Societies in 2002. These three- to six-month assignments provide interns with first-hand experience working in national science policy. They work closely with the Director of Science Policy, gathering information, attending briefings, and preparing position statements. The Washington Office of Science Policy has enabled the Societies to provide scientific input to policy development and has added an important aspect to the Societies' role during the first decade of the 21st century. It is a high priority of the Society.

RELATIONSHIP WITH OTHER SOCIETIES

Throughout its existence ASA has recognized the value of cooperation and participation (Laude, 1962). The ASA itself came into existence at the meeting of the American Association for the Advancement of Science in 1907. In 1910, the title of A.M. Ten Eyck's (third president) presidential address (Ten Eyck, 1911) was "The Affiliation of American Agricultural Societies." Throughout the years cooperation and affiliation with other scientific societies has been a priority. There have been numerous joint meetings with other societies, as well as various levels of liaisons with other societies. The Societies were founding members of the Council for Agricultural Science and Technology in 1973 (Smith, 1980). In the last 10 years there has been additional emphasis on cooperation with other professional societies, such as the European Society of Agronomy, Canadian Society of Agronomy, and American Institute of Biological Sciences. Officers or other members frequently attend meet-

Bringing science to Washingtion, a delegation of Congressional Science Fellows, including Lee van Wychen, the ASA Fellow, visits Capitol Hill in 2003.

ings of other societies and organizations. Likewise, other societies often send representatives to ASA annual meetings.

AGRONOMIC SCIENCE FOUNDATION

The Agronomic Science Foundation (ASF) was founded in 1967 as a 501(c)(3) nonprofit charitable foundation and began operations in 1968. Articles of incorporation are published in *Agronomy Journal* 60:133–135. The mission as defined in the original Articles of Incorporation is to organize the foundation solely for scientific, educational, and charitable purposes in the broad interest of agronomy, crop science, soil science, and other related disciplines. Today's mission is "To provide leadership and financial resources to further the role of the agronomic, crop, and soil sciences in global crop production, and to promote human welfare within a sustainable environment."

The Board of Trustees has governed ASF during its 40-year history. They elected a President and Vice President from inception until 2000. Since then the officers titles have changed to Chair and Vice-Chair. The Executive Vice President of the Societies has been the Secretary-Treasurer. The first paid director of ASF activities, John Kruse, was hired in 1998. Most of the ASF funds have been established to support specific activities such as awards, lectureships, or scholarships. From $26 in 1969, the funds in the ASF investment portfolio have grown to $2.7 million by the end of 2006. Total assets in cash and investments were $3.1 million. From 1989 through 2006, ASF has provided a total of $883,680 in support of Society activities.

During 2007, ASF is undergoing significant structural changes and is redesigning its operational procedures. In the future ASF will be called upon by the Societies to continue to seek funds for support of activities and special projects. Many of these projects likely will require large amounts of funds. In the last several years ASF has worked with SSSA to raise funds for a special Smithsonian National Museum of Natural History soils exhibit and with the CSSA for the Golden Opportunity Scholars Program (Cal Qualset, ASF Board of Directors, 2007, personal communication).

LOOKING TO THE FUTURE

As the first century of ASA comes to a close and we embark on the second, our organization has grown from 43 interested people to three independent Societies, international in scope, that operate cooperatively, with a Society headquarters employing 43 people. ASA reached a membership peak in 1986 with 12,781 members. As of 2007, there are about 8000 members. There are numerous reasons for the decline. Two key reasons are increased specialization of scientists and the change that eliminated automatic ASA membership for CSSA and SSSA members. The three Societies have just completed major structural change changes that put all three Societies in a parallel position, each as an independent Society. The ASA now has a much smaller board of directors that can concentrate on programming for ASA. ASA has experienced major structural changes in 1931, 1947, and 2006, as well as in 1936, when SSSA was formed and 1955, when CSSA was formed. Each time ASA has developed strong, responsive programs. This challenge faces us again today as each Society becomes separate and independent. However, with the new structure great opportunities exist for the Societies, who, although independent, can operate collaboratively to meet crop, soil, and environmental challenges that are likely to intensify in the future.

At this juncture of the first and second centuries for ASA, President Jerry L. Hatfield and the ASA board of directors look to the future and present the following statement:

There have been many changes in ASA since it was formed 100 years ago; however, over the next few years there may be some of the largest changes in the Society. The restructuring of the ASA board of directors and engagement of the board in all decisions for the Society represent the beginning of a series of changes that will position ASA to remain as a leader in scientific societies. The formation of a management unit for the Societies, the Alliance of Crop, Soil, and Environmental Science Societies, will allow the ASA executives and board to focus more closely on their function as a professional society than as a management entity for multiple societies. In May 2006, the ASA Board developed a strategic plan that contained some aggressive goals. These goals are:

Goal A: ASA will be the members' indispensable resource for leading-edge education, knowledge, and networking.
Goal B: ASA will be recognized as the powerful advocate and voice for advancing the science of agronomy to proactively address emerging global issues.
Goal C: The integrated science of agronomy will be recognized by key audiences as the source of science-based knowledge that improves the management of soils, crops, and the environment.
Goal D: ASA will be known for its innovative infrastructure that advances and sustains its success.

These are aggressive goals that will require ASA to develop strong partnerships across the Society that extend from discovery through delivery. Discovery will not only include research but also new methods of education. Delivery will encompass the Certified Crop Advisers linked with the research and education efforts. This continuum will require new approaches, new delivery methods, and new interactions among all sectors of ASA. An example of broadening of the ASA scope is the development of interaction of ASA with the European Society of Agronomy and the International Environmental Modeling and Simulation Society to develop a series of Farming System Design workshops to promote international exchanges on agronomic questions. These are exciting changes that are underway and will position ASA to retain its leadership status that was envisioned 100 years ago when the Society was formed.

The ASA objective statement at its founding, 100 years ago, "The increase and dissemination of knowledge of crops and soils and the conditions that affect them" fits our mission fairly well as we enter the second century. Today, we, likely, would add "...and how they affect the environment" to that original mission statement.

REFERENCES

ASA. 1910. Business section. Proc. Am. Soc. Agron. 1:6–15.
ASA. 1931. Agronomic affairs. J. Am. Soc. Agron. 23:1020–1078.
ASA. 1946. Minutes of the thirty-eighth annual meeting of the Society. J. Am. Soc. Agron. 38:1119–1130.
ASA. 1951. Agronomic affairs. Agron. J. 43:48–56.
Black, C.A., C.H. Hansen, and R.J. McCracken. 1972. Foreword. J. Environ. Qual. 1:iii.
Carleton, M.A. 1910. Development and proper status of agronomy. Proc. Am. Soc. Agron. 1:17–23.
Fuccillo, D.A. 1983. The 75th publication year of *Agronomy Journal*. Agron. J. 75:413–417.
Gast, R.G., D.N. Duvick, and L.L. Boersma. 1988. Foreword. J. Prod. Agric. 1:1.
Keim, F.D. 1953. History of the American Society of Agronomy (1941–1953). Agron. J. 45:651–653.
Laude, H.H. 1962. History of the American Society of Agronomy, First fifty years—1907–1957. Agron. J. 54:57–69.
Luellen, W.R. 1987. We bid you adieu. Crops and Soils 39(9):35–38.
Luckett, J.D. 1962. Publications: The first forty years. p. 59–62. *In* H.H. Laude. History of the American Society of Agronomy. First fifty years—1907–1957. Agron. J. 54:57–69.
Lyon, T.L. 1933. History of the organization of the American Society of Agronomy. J. Am. Soc. Agron. 25:1–9.
McFee, B. 1999. A unique partnership: ASA, CSSA, and SSSA. Crop Science–Soil Science–Agronomy News, March 1999, p. 11–12.
Metcalfe, D.S. 1962. Student activities sponsored by the society. p. 64–66. *In* H.H. Laude. History of the American Society of Agronomy. First fifty years—1907–1957. Agron. J. 54:57–69.
Milford, M.H., and D.T. Smith. 2007. Development and evolution of resident education in the American Society of Agronomy. p. 69–78. *In* L.E. Moser (ed.) The American Society of Agronomy: 100 years of history. ASA, Madison, WI.
Miller, M.F. 1962. Founding of the American Society of Agronomy. p. 57–59. *In* H.H. Laude. History of the American Society of Agronomy. First fifty years—1907–1957. Agron. J. 54:57–69.
Pearson, C.H., S.M. Ernst, K.A. Barbarick, J.L. Hatfield, G.A. Peterson, and D.R. Buxton. 2007. *Agronomy Journal* turns one hundred. p. 59–68. *In* L.E. Moser (ed.) The American Society of Agronomy: 100 years of history. ASA, Madison, WI.
Rohweder, D.A., R.F Barnes, D.M. Kral, and C.O. Qualset. 2007. Development of the American Society of Agronomy, 1978–2007. p. 47–68. *In* L.E. Moser (ed.) The American Society of Agronomy: 100 years of history. ASA, Madison, WI.
Smith, D. 2007. Building the ASA headquarters in Madison. p. 45–46. *In* L.E. Moser (ed.) The American Society of Agronomy: 100 years of history. ASA, Madison, WI.
Smith, D.C. 1980. Development of the American Society of Agronomy, 1958–1977. Agron. J. 72:227–240.
Ten Eyck, E.M. 1911. The affiliation of American agricultural societies. Proc. Am. Soc. Agron. 2:33–35.

Mark Carleton

Mark Alfred Carleton
President, American Society of Agronomy 1907–1908

Mark Alfred Carleton was born March 7, 1866, near Jerusalem, in Monroe County, Ohio, the son of Louis D. Carleton, whose father, Abner G. Carleton, of English descent, migrated to Ohio from Pennsylvania. His mother, whose maiden name was Lydia J. Mann, is of Dutch ancestry. There are now a number of Carleton families in Pennsylvania and in Maryland near Washington who appear, from certain indications, to be descended from the same ancestors of a century and a half ago.

In 1876, when he was ten years of age, his family moved to a farm in Cloud County, Kansas. His early education was obtained in the rural schools of Ohio and Kansas. In 1884 he entered the sophomore class of the Kansas Agricultural College at Manhattan, completing his course and also a year of special work in biology and chemistry, and graduating with the degree of Bachelor of Science in 1887. He became Professor of Natural History in Garfield University (now Friends University) at Wichita, Kans., during 1890–91. During 1891–92 he taught natural history in Wichita University, and during 1892–93 took a post-graduate course in botany and horticulture at the Kansas Agricultural College, receiving the degree of Master of Science. During 1893 he was Assistant Botanist at the Kansas Experiment Station, his time being devoted chiefly to plant pathology and particularly to the rusts of cereals. While teaching in Wichita, three years of Latin and one year of Greek were taken under private teachers.

In March, 1894, Professor Carleton began his service in the United States Department of Agriculture by appointment as Assistant Pathologist in the Division of Vegetable Physiology and Pathology, giving special attention to cereal diseases. During his seven years of service in pathological work he established the physiological relationships of nearly all the cereal rusts of this country and demonstrated the distinctness of the different forms of the same species of these rusts adapted to the same hosts, traced the yearly cycle of the orange leaf-rust of wheat, and showed that durum wheats, emmers, einkorns, and some other wheats are more or less resistant to rust.

Since 1901 he has been Cerealist in Charge of Grain Investigations in the Bureau of Plant Industry. In this position his work has included the introduction of new varieties adapted to this country; thorough trials at numerous experiment farms of these varieties and others produced by hybridization and selection; the breeding of small grains, about one hundred hybrids being now under experiment; direction of field and chemical experiments to determine the effect of environment on the composition of cereals; the direction of investigations in cereal diseases; personal studies of the characteristics of numerous varieties of wheat, oats and grain millets; study of practical farm methods in cultivation, rotations, etc.; particularly in dry-land districts; taking part in grain expositions, farmers' institutes, agricultural congresses, judging exhibits, etc.

Some of the more permanent results of his cereal investigations have been the introduction of durum wheat into this country where it is now an established crop, yielding 60,000,000 bushels of wheat annually; the establishing of the Swedish Select oat, which now furnishes

Mark Alfred Carleton, President, 1907–1908.

Left: Wheat explorer Mark Carleton, wearing black hat, expands his tireless research efforts from the Great Plains of America to the Steppes of Russia. This sketch of Carleton meeting with Russian farmers appears in a 1926 issue of Country Gentleman, *accompanying the article "Carleton The Wheat Hunter." The sketch is captioned, "Then Carleton Wandered in His Fantastic Hunt for Wheat Up and Down the Black Earth of Russia." Reprinted from* The Country Gentleman *magazine, copyright 1926 Saturday Evening Post Society. www.saturdayeveningpost.com.*

Section page: The study of wheat was Carleton's life-long work, and his influence on modern cultivars is evident today.

40,000,000 bushels of the annual oat crop; the direction of investigations establishing the Sixty Day oat, now the most popular variety as a "general purpose" oat in this country; the introduction of hardier strains of the Turkey or Crimean group of wheats, including the Kharkov, which yields now about 10,000,000 bushels of the wheat crop of this country; the introduction of Black Winter emmer, a very hardy cereal for stock food; introduction of the cultivation of winter barley into the Middle Western States, thus permitting a large increase in the acre-yields of the barley crop; experiments showing the pronounced effect of the presence of water in the deterioration of the gluten content of wheat, and the inauguration and direction of experiments in the Texas Panhandle, which have had much effect in establishing rational dry farming.

During 1898 and 1899 he was an Agricultural Explorer in eastern Europe and Siberia, in search of rust-resisting and drought-resisting cereals. In 1900 he was Expert in charge of the grain exhibit of the United States at the Paris Exposition. In the same year he was reappointed as Agricultural Explorer for another trip in eastern Europe in search of hardy cereals and to increase the supply of those originally obtained. In 1904 he was Chairman of Group VIII of the International Jury at the Louisiana Purchase Exposition, St. Louis.

On December 29, 1897, he was married to Amanda Elizabeth Faught, who was born at Kingman, Kans., in 1874, the daughter of R.D. and Hannah Faught.

Professor Carleton took an active part in the work preliminary to the organization of this Society. Probably more to him than to any other one person belongs the credit of founding it. At the first meeting held in Chicago, December 31, 1907, to January 1, 1908, he was unanimously chosen the first president of the infant Society. During these formative first years of its history he had a very large share in determining its growth and development.

The chief publications from the pen of Professor Carleton are listed below.

PUBLICATIONS

Bulletins of Kansas Experiment Station
Preliminary Report on Rusts of Grain (with Hitchcock). Bul. 38. 1893.
Rusts of Grains II (with Hitchcock). Bul. 46. 1894.

Publications in the U. S. Department of Agriculture

Division Vegetable Physiology and Pathology:
Cereal Rusts in the United States. A Physiological Investigation. Bul. 16. 1899.
Basis for the Improvement of American Wheats. Bul. 24. 1900.
A New Wheat Industry for the Semiarid West. Cir. 18. 1901.

Division of Botany:
Russian Cereals Adapted for Cultivation in the United States. Bul. 23. 1900.

Bureau of Plant Industry:
Macaroni Wheats. Bul. 3. 1901.
Investigations of Rusts. Bul. 63. 1904.
The Commercial Status of Durum Wheat (with Chamberlain). Bul. 70. 1904.
Ten Years' Experience with the Swedish Select Oat. Bul. 182. 1910.
Barley Culture in the Northern Great Plains. Cir. 5. 1908.

Farmers' Bulletins:
Emmer: A Grain for the Semiarid Regions. 139. 1901.
Lessons from the Grain-Rust Epidemic of 1904. 219. 1905.

Yearbook Separates.
Improvements in Wheat Culture. 1896:489–498.
Successful Wheat-Growing in the Semiarid Districts. 1900: 529–542.
The Future Wheat Supply of the United States. 1909: 259–272.

Principal Contributions to Journals
Second List of Kansas Parasitic Fungi (with Kellerman). Trans. Kan. Acad. Sci. X:88–99. 1886.
Characteristic Sand Hill Flora. Ibid. XII:32–34. 1889.
Variations in Dominant Species of Plants. Ibid. XIII:24–28. 1891.
List of Plants Collected by the Garfield University Expedition of 1889. Ibid. XIII:50–57. 1891.
Observations on the Native Plants of Oklahoma Territory and Adjacent Districts. Contr. Nat. Herb. I, No. 6: 220–232. 1892.
Notes on the Occurrence and Distribution of Uredineae. Science XXII:62–63. 1893.
Studies in the Biology of the Uredineae I—Notes on Germination. Bot. Gaz. XVIII:447–457. 1893.
Millets. Bailey's Cycl. Amer. Agric. Vol. II, Crops. pp. 469–474.
Report on Vegetable Food Products, Class 39, Paris Exposition of 1900, in Rep. Com.-Gen. for U.S. to Paris Exposition, Vol. V:314–321.
Culture Methods with Uredineae. Jour. Appl. Micros. & Lab. Meth. VI, No. 1:2109–2111. 1902.
Development and Proper Status of Agronomy. Proc. Am. Soc. Agron. 1:17–23. 1908–09.
Limitations in Field Experiments. Proc. Soc. Prom. Agric. Sci. 1909: pp. 55–61. 1910.
The Future Wheat Supply of the United States. Science XXXII: pp. 161–171. 1910.

The Agronomic Legacy of Mark A. Carleton

Gary M. Paulsen
Department of Agronomy
Kansas State University., Manhattan, Kansas

Mark A. Carleton is known by members as the first president of the American Society of Agronomy. However, he also shaped development of the hard red winter wheat (*Triticum aestivum* L.) and durum wheat (*T. turgidum* L. var. *durum*) industries in the USA. The objective of this article is to describe Carleton's agronomic legacy: introduction of adapted wheat cultivars, dryland production of wheat, investigations of wheat diseases, and scientific knowledge of wheat. Carleton was the first scientist to recognize the superiority of 'Turkey' hard red winter wheat, which came to dominate the central and southern Great Plains; introduced 'Kharkof' wheat, which became an important cultivar in the central and northern Plains; introduced 'Crimean' wheat, which was a parent of many early improved cultivars; and introduced 'Kubanka' and promoted its production to start the durum wheat industry in the northern Great Plains. He gave the common name to leaf rust (caused by *Puccinia recondita* f. sp. *tritici*), determined the physiological relationships among known races of leaf rust and stem rust (caused by *P. graminis* f. sp. *tritici*), and was among the first to investigate fungicides for controlling rust diseases. Research directed by Carleton developed suitable methods of dryland farming for the Great Plains and established the relationship between soil moisture and gluten quality of wheat. His publications were a blueprint for development of the U.S. wheat industry, and his book, *The Small Grains*, became a classic.

Members of the American Society of Agronomy should be aware of Carleton's achievements because he was considered the most respected agronomist in the USA during his career (Isern, 2000), and he defined the profession in principle and in practice (e.g., Carleton, 1907–1909, 1915a). It is particularly appropriate to recognize his agronomic legacy because of the controversy that surrounded latter aspects of Carleton's life (de Kruif, 1928).

INTRODUCTION TO THE LEGACY

Carleton was born in Ohio in 1866, reared in Kansas, and earned a B.S. degree in agriculture from Kansas State Agricultural College (KSAC) (now Kansas State University) in 1887. After teaching at Garfield University (now Friends University) at Wichita from 1888 to 1890, he returned to KSAC for an M.S. degree, which was conferred in 1893. He then was assistant botanist at KSAC until 1894, when he transferred to the USDA and served his most productive years as cerealist and chief of cereal investigations.

A small group of persons, including Carleton, met at the University of Chicago on 31 Dec. 1907, to form the American Society of Agronomy (Slate, 1952). Carleton had conducted much of the correspondence that led to organization of the Society and was unanimously selected as its first president (Ball and Warburton, 1925). The early growth and development of the Society were greatly influenced by him. An extraordinary

Series of hundredth-acre wheat plots in foreground, with oat plots in the background, as pictured in the 1910 volume of Proceedings of the American Society of Agronomy. *Carleton's contributions to the study of wheat changed the face of American agriculture.*

Reprinted from
J. Nat. Resour. Life Sci. Educ., 30:120–123, 2001

Corresponding author
G.M. Paulsen, Emeritus Professor, Crop Physiology, 3717 Throckmorton, Department of Agronomy, Kansas State Univ., Manhattan, KS 66506-5501 (email: gmpaul@ksu.edu).

Abbreviations: KSAC, Kansas State Agricultural College; KAES, Kansas Agricultural Experiment Station.

agronomic legacy was left by Carleton in addition to his contributions to the American Society of Agronomy. Carleton, more than any other person, shaped the early hard red winter wheat and durum wheat industries in the USA. As noted by Parker (1935), "Hard red winter wheats are a monument to the far-sightedness of M.A. Carleton in agronomy, plant pathology, and plant breeding," and according to Salmon et al. (1953), "The early development of the (durum) industry was due largely to the initiative and vision of M.A. Carleton." Ball (1948) concluded that "To Carleton goes the credit due the discoverer, the pioneer who made our wheat industry what it is today."

BACKGROUND OF THE LEGACY

An understanding of Carleton's contributions must be placed in the context of Great Plains agriculture during the late 19th and early 20th centuries. Settlers from the eastern USA and western Europe had homesteaded the region, bringing the seeds and technology for growing crops from their native lands. Many types of wheat were introduced, most of them ill-adapted to the environment, and the settlers' familiar farming practices were mostly inappropriate for Great Plains conditions. Spring wheat predominated in the region, and soft wheat was favored because of its ease of milling (Heyne, 1987).

Two important advances in production of wheat in the central and southern Great Plains occurred during the 1870s. Turkey hard red winter wheat from the Crimea was introduced to central Kansas by Mennonite settlers from the Ukraine in 1873. The new cultivar was well adapted—winter hardy, drought resistant, and produced excellent quality grain for baked products. The invention of the steel roller mill in 1878 facilitated milling of the new cultivar and eliminated widespread discounts on hard grain (Heyne, 1987).

Durum wheat was a minor crop in the northern Great Plains at the turn of the century, with only about 2500 Mg of grain produced in 1901 (Salmon et al., 1953). The only cultivar that was grown commercially was 'Amautka', which was brought to the Dakotas by immigrants from Russia sometime before 1898 (Clark, 1936).

Technology for wheat production was labor-intensive and crop failures were frequent. Information was particularly needed on seedbed preparation, conservation of soil moisture, and control of pests. However, "A farmer was much more likely to consult his almanac than the experiment station to determine when to sow wheat..." (Salmon et al., 1953).

A LEGACY OF HARD RED WINTER WHEAT

Initial spread of Turkey wheat in the Great Plains was slow, because the Mennonite community in Kansas was close-knit and seed of the new cultivar was scarce (Heyne, 1987). Carleton was widely acknowledged as the first scientist to recognize the advantage of Turkey wheat (Ball, 1930, 1948; Clark, 1936). He probably became acquainted with the cultivar when he was at Garfield University, which is near the area where the Mennonites settled. Carleton had a keen eye for variation in plants and led university expeditions to collect specimens in the area (Carleton, 1891–1892a, 1891–1892b). However, he never claimed credit for identifying the value of Turkey, stating only that "...its merits did not become generally known until about 1890" (Carleton, 1916).

The Kansas Agricultural Experiment Station (KAES) compared Turkey with several popular soft red winter wheat cultivars and reported it as "...coming to the front as a heavy yielder" and "...perhaps the hardiest wheat of any we have tested" (Georgeson et al., 1896). Official recognition of its superiority and serious freeze damage to other cultivars during the late 1890s prompted general acceptance of Turkey. By 1919, when the first official cultivar survey was made, Turkey occupied more than 82% of the Kansas wheat area. It remained the most popular cultivar in the state until 1939 and in the USA until 1944. Today, hard red winter wheat is the most important class in the USA, being grown on more than 40% of the country's wheat area.

The excellent adaptation of Turkey to the Great Plains motivated Carleton to seek other useful cultivars in Russia (Carleton, 1897). In expeditions to there and Siberia for the USDA during 1898–1899 and 1900, he returned with many new crops, including Kharkof and Crimean hard red winter wheat and Kubanka durum wheat (Carleton, 1900a). Carleton established ties with state agricultural experiment stations to test the new cultivars, a course that was unprecedented in the USDA (Isern, 2000). Kharkof became a popular cultivar in its own right in the Great Plains. Official surveys are unavailable, however, because farmers and statisticians commonly labeled all cultivars from Russia as Turkey. Estimates of the area credited to Turkey that was actually occupied by Kharkof range from about half of the Kansas wheat crop in 1914 (Carleton, 1915a) to 20% in 1919 (Anonymous, 1920). Kharkof also expanded the hard red winter wheat region into Nebraska and Montana, where spring wheat predominated into the 1900s, because of its greater winter hardiness than Turkey (Salmon et al., 1953). According to Carleton (1915a), Montana "...has practically been made a wheat state by the use of Kharkof...", and by 1919, "...by far the greater part" of the U.S. crop of hard red winter wheat was sown with Kharkof (de Kruif, 1928). Ball (1948) considered Kharkof "...the greatest import this country has ever enjoyed."

Crimean became an important cultivar in the Great Plains and excelled as a parent of other cultivars. 'Kanred', the first improved cultivar released by the KAES (in 1917), and 'Cheyenne', an early, important cultivar developed by the Nebraska Agricultural Experiment Station (in 1930), were direct selections from Crimean. A selection from Kanred, P-1066, was a parent of 'Tenmarq', the first improved cultivar from hybridization released by the KAES (in 1932).

A LEGACY OF DURUM WHEAT

Annual production of durum wheat in the USA increased from its low point in 1901 to about 1,850,000 Mg on 2,100,000 ha in North Dakota and surrounding states during 1925–1929 (Salmon et al., 1953). This dramatic increase was due almost solely to the initiative of Carleton (Ball, 1930; Clark, 1936). In addition to introducing Kubanka in 1900, he also brought additional supplies of Arnautka from Russia, increased and distributed seed of both cultivars in the Dakotas, and promoted production of the new wheat class by farmers (Carleton, 1901a; Salmon et al., 1953). Kubanka became the most popular durum cultivar and the standard for rating the quality of all other cultivars (Salmon et al., 1953). The cultivar was so popular that it was usually known as durum wheat, and it continued to dominate production into the 1940s (Joppa, 1988).

Kubanka was as important for developing improved cultivars as for producing grain. 'Nudak' from North Dakota, 'Mondak' from North Dakota and Montana, and 'Acme' from South Dakota were direct selections from Kubanka. One of the first improved cultivars of durum wheat developed by hybridization by the North Dakota Agricultural Experiment Station was named after Carleton in 1943.

A LEGACY OF DISEASES OF WHEAT

Carleton's abiding interest was stem rust of wheat. His early investigations also concerned leaf rust, and Chester (1946) credited Carleton (1899) with coining the common name for the disease. However, Carleton was not convinced that the disease was a problem.

According to legend, Carleton began studying the rust diseases on his parent's farm at the age of 11 and was a self-trained botanist when he entered college (Ball, 1948). The thesis for his M.S. degree concerned germination of rust spores (Carleton, 1893a, 1893b), and his early work as a scientist was among the first in the USA to evaluate fungicides for control of rust diseases on cereals (Hitchcock and Carleton, 1893, 1894).

Carleton's greatest contribution to knowledge of rust diseases was in determining their physiological relationships. He was among the early workers who noted that cereal cultivars were resistant to rust diseases in some countries and susceptible in other countries, a phenomenon that was previously attributed to differences in the constitution of plants from changes in the climate (Chester, 1946). Concurrent with scientists in Sweden, Hitchcock and Carleton (1894) concluded that their experiments "...seem to show that the rusts of various cereals are probably physiological species," which was among the first recognition that differences in susceptibility were due to specialization of the causal organism. After establishing the physiological relationships among most of the rust diseases in the USA, Carleton went on to demonstrate that cereal cultivars reacted differently to the fungi and ranged in susceptibility to them (Carleton, 1899). His work with rust diseases made Carleton "...the leading plant pathologist of America" (de Kruif, 1928).

The work of Carleton with rusts contributed greatly to understanding host–parasite relationships and improving disease resistance of cereals. However, some of his inferences were erroneous. He considered leaf rust to be benign and even beneficial because preventing excessive foliage aided growth of the grain (Carleton, 1899). His influence caused other scientists to feel that they were "...wasting much time attending to the spotting rust" in their breeding programs (Chester, 1946). Carleton also believed that the transition from spring wheat to winter wheat had largely eliminated the problems of rust diseases in the Great Plains, and he advocated that only one cultivar should be grown in a locality or state (Carleton, 1915b). Today, of course, rusts continue to be important diseases, and genetic vulnerability to new races is a major cause of the profusion of wheat cultivars.

A LEGACY OF PRODUCING AND PROCESSING WHEAT

Technology for producing and processing wheat was as deficient as suitable cultivars for the Great Plains during the late 19th century. In an early publication, Carleton (1897) realized the limitations of soil moisture and advocated deep, early plowing; fine, mellow seedbed; rough soil surface; and early sowing of appropriate cultivars. He also recommended the new feature of hybridization to add vigor to self-pollinated wheat and introduction of new seed to counter running out of cultivars.

His expeditions to Russia to collect seeds of new crops also familiarized Carleton with production conditions that were even more extreme than in the Great Plains. Peasants in Russia, through long practice, had learned to cope with drought, harsh temperatures, and other adversities, and those experiences made them especially successful at growing wheat when they immigrated to the USA. Carleton (1900a) made detailed notes of the climate in Russia and Siberia, where he collected seeds, and of the methods used by farmers to till the soil, sow the seed, and harvest the grain.

Carleton applied his experiences in Russia to improving wheat production in the USA when he was named chief of cereal investigations by the USDA in 1901. The Bureau of Plant Industry of the USDA initiated a Dryland Agriculture Program, and Carleton directed much of the research on farming methods in semiarid areas of the USA (Carleton, 1901c, 1915b). The recognition of the relationship between soil moisture and gluten content and quality of wheat was an important result (Ball and Warburton, 1925).

Establishing durum as a new crop presented several problems (Carleton, 1901b). Besides increasing and distributing seed of adapted cultivars, farmers in the Dakotas had to be convinced to grow the new class of wheat. The campaign was facilitated by Carleton's vigorous promotion of durum and failure of the hard red spring wheat in the region from an epidemic of stem rust (Carleton, 1905; Clark, 1936). Carleton also directed extensive milling and baking tests to determine the quality of durum produced in the northern Plains (Carleton and Chamberlain, 1904). However, millers, who were accustomed to producing flour for pasta from common wheat, had to be persuaded to accept the extremely hard durum (Carleton, 1915a). Early harvests were mostly exported to Europe and some even went to Russia in 1905. By 1911, however, domestic millers had largely accepted durum, even at premium prices.

A LEGACY OF KNOWLEDGE OF WHEAT

Four publications among those authored by Carleton stand out. The first paper reported the physiological relationships among the known races of stem rust (Carleton, 1899). The second publication was termed a *foundation paper* by Ball (1930), because it provided the basis for much of the subsequent development of the U.S. wheat industry. In the paper, Carleton (1900b) summarized his evaluation of the resistance of nearly 1000 cultivars of wheat to rust, drought, and cold as well as their grain quality traits. He also analyzed the characteristics of the different regions of the U.S. for producing wheat, the classes of wheat that were suited for each region, and the cultivar traits that were required for adaptation of each class. The analysis, in retrospect, was a blueprint for much of Carleton's subsequent efforts with wheat.

Although it appeared 3 yr before Carleton's career with wheat ended, an article in the 1914 *Yearbook of the USDA* was an epilogue of his accomplishments (Carleton, 1915a). The article described the development of the hard wheat industries in the USA—the introduction, adaptation, and characteristics of important cultivars; acceptance of the different classes by millers; and the impact on national production and marketing. Of the three classes of hard wheat in the USA (hard red spring, hard red winter, durum), Carleton had played a major role in two of them that accounted for more than half of the wheat in the country, and cultivars identified by him (Turkey, Kharkof, Kubanka) dominated production of those classes.

One of Carleton's final publications was his tome, *The Small Grains* (Carleton, 1916). A review called the book "interesting...commendable... very good" and "suitable for...colleges and every cereal farmer's library" (Hayes and Olson, 1917). The book was the most comprehensive and current text of its time, covering the major small-grain cereals and their morphology, nutrition, improvement, adaptation, cultivation, pests, and uses. His immediate experience and national perspective made Carleton uniquely qualified to author the book.

IMPACT OF THE LEGACY

Mark A. Carleton, as an organizer and first president of the American Society of Agronomy, influenced generations of agronomists. His accomplishments as an agronomist also benefited many generations of other persons—wheat growers, millers, bakers, and consumers. One must agree with Parker that "Grain growers, the grain trade, and the grain processing industry in the United States owe an incalculable debt of gratitude to Mark Alfred Carleton" (Swanson, 1958).

The legacy of Carleton's contributions persists today. Hard red winter wheat and durum wheat are still major crops in the Great Plains and main sources of livelihoods for farmers, their families, and their communities. Modern cultivars of hard red winter wheat are greatly changed, but half of their genes still trace to Turkey and Kharkof (Cox, 1991). The durum wheat industry, which was largely created by Carleton, remains an important enterprise in the northern Great Plains. Kubanka, the original cultivar, has been displaced but is still the foundation of modern durums and the standard for rating their quality (Joppa, 1988).

Other crop introductions by Carleton had mixed success. 'Swedish Select' oat (*Avena sativa* L.), which was introduced from Russia in

1898 (Carleton, 1910), became the most popular cultivar in the Upper Midwest but diminished in importance as tractors replaced horses on farms (Isern, 2000). Emmer wheat (*T. dicoccum* Schrank), which was promoted as a feed grain, never caught on because other species were more productive (Carleton, 1901a; Isern, 2000).

Carleton did not share in the benefits of his legacy. Pyramiding debts to colleagues and grain dealers that began with the death of one child and hospitalization of another created a scandal that led to his resignation from the USDA in 1918 (Isern, 2000). His family lost their home and, brokenhearted, he drifted to a series of minor agricultural posts in Central and South America (Ball, 1948). He struggled to repay his obligations but died of heart disease exacerbated by acute malaria in Peru in 1925,"...most miserably neglected" and "...among almost total strangers, far from his family, friends, and native land" (de Kruif, 1928; Swanson, 1958). He is interred in the small village of Paita in Peru, with "...no memorial...proclaiming what he did for his home country" (Isern, 2000).

Acknowledgment

Appreciation is expressed to the staff of the Kansas State University Archives for providing historical information.

REFERENCES

Anonymous. 1920. Analysis of data on wheat production in Kansas. p. 16–138. *In* Wheat in Kansas. Kansas State Board of Agric., Topeka, KS.

Ball, C.R. 1930. The history of American wheat improvement. Agric. Hist. 4:48–71.

Ball, C.R., and C.W. Warburton. 1925. Mark Alfred Carleton. J. Am. Soc. Agron. 17:514–516.

Ball, J.W. 1948. Grain hunters of the west. Nation's Business 36(9):5440.

Carleton, M.A. 1891–1892a. Variations in dominant species of plants. Trans. Kansas Acad. Sci. 13:24–28.

Carleton, M.A. 1891–1892b. List of plants collected by the Garfield University expedition of 1889. Trans. Kansas Acad. Sci. 1350–57.

Carleton, M.A. 1893a. Studies in the germination of *Uredineae*. M.S. thesis. Kansas State Agric. College, Manhattan, KS.

Carleton, M.A. 1893b. Studies in the biology of *Uredineae*: I. Notes on germination. Bot. Gaz. 18:47–457.

Carleton, M.A. 1897. Improvements in wheat culture. p. 489–498. *In* 1896 Yearbook of the USDA. U.S. Gov. Print. Office, Washington, DC.

Carleton, M.A. 1899. Cereal rusts of the United States: A physiological investigation. USDA Div. Veg. Phys. Pathol. Bull. 16. U.S. Gov. Print. Office, Washington, DC.

Carleton, M.A. 1900a. Russian cereals adapted for cultivation in the United States. USDA Div. Bot. Bull. 23. U.S. Gov. Print. Office, Washington, DC.

Carleton, M.A. 1900b. The basis for the improvement of American wheats. USDA Div. Veg. Phys. Pathol. Bull. 24. U.S. Gov. Print. Office, Washington, DC.

Carleton, M.A. 1901a. Emmer: A grain for the semiarid regions. USDA Farmers' Bull. 139. U.S. Gov. Print. Office, Washington, DC.

Carleton, M.A. 1901b. Macaroni wheats. USDA Bur. Plant Ind. Bull. 3. U.S. Gov. Print. Office, Washington, DC.

Carleton, M.A. 1901c. Successful wheat growing in semi-arid districts. p. 529–542. *In* 1900 Yearbook of the USDA. U.S. Gov. Print. Office, Washington, DC.

Carleton, M.A. 1905. Lessons from the grain rust epidemic of 1904. USDA Farmers' Bull. 219. U.S. Gov. Print. Office, Washington, DC.

Carleton, M.A. 1907-1909. Development and proper status of agronomy. J. Am. Soc. Agron. 1:17–23.

Carleton, M.A. 1910. Ten years' experience with the Swedish Select oat. USDA Bur. Plant Ind. Bull. 182. U.S. Gov. Print. Office, Washington, DC.

Carleton, M.A. 1915a. Hard wheats winning their way. p. 391–420. *In* 1914 Yearbook of the USDA. U.S. Gov. Print. Office, Washington, DC.

Carleton, M.A. 1915b. Problems of the wheat crop. J. Am. Soc. Agron. 7:78–84.

Carleton, M.A. 1916. The small grains. The MacMillan Co., New York, NY.

Carleton, M.A., and J.S. Chamberlain. 1904. The commercial status of durum wheat. USDA Bur. Plant Ind. Bull. 70. U.S. Gov. Print. Office, Washington, DC.

Chester, K.S. 1946. The cereal rusts. Chronica Botanica Co., Waltharn, MA.

Clark, J.A. 1936. Improvement in wheat. p. 207–302. *In* 1936 Yearbook of the USDA. U.S. Gov. Print. Office, Washington, DC.

Cox, T.S. 1991. The contribution of introduced germplasm to the development of U.S. wheat cultivars. p. 25–47. *In* Use of plant introductions in cultivar development. CSSA Spec. Publ. 17. CSSA, Madison, WI.

de Kruif, P. 1928. The wheat dreamer Carleton. p. 2–30. *In* Hunger fighters. Harcourt, Brace & World, New York, NY.

Georgeson, C.C., F.C. Burtis, and D.H. Otis. 1896. Experiments with wheat. Kansas Agric. Exp. Stn. Bull. 59.

Hayes, H.K., and P.J. Olson. 1917. New books: The small grains. J. Am. Soc. Agron. 9:196198.

Heyne, E.G. 1987. The development of wheat in Kansas. p. 41–56. *In* G.E. Ham and R. Higham (ed.) The rise of the wheat state. Sunflower Press, Manhattan, KS.

Hitchcock, A.S., and M.A. Carleton. 1893. Preliminary report on rusts of grain. Kansas Agric. Exp. Stn. Bull. 38.

Hitchcock, A.S., and M.A. Carleton. 1894. Second report on rusts of grain. Kansas Agric. Exp. Sm. Bull. 46.

Isern, T.D. 2000. Wheat explorer the world over: Mark Carleton of Kansas. Kansas Hist. 23: 12-25.

Joppa, L.R. 1988. Genetics and breeding of durum wheat in the United States. p. 47–68. *In* G. Fabriana and C. Lintas (ed.) Durum wheat chemistry and technology. Am. Assoc. of Cereal Chemists, St. Paul, MN.

Parker, J.H. 1935. Wheat improvement in Kansas, 1874–1934. p. 139–164. *In* Twenty-ninth biennial report. Kansas State Board of Agric., Topeka, KS.

Salmon, S.C., O.R. Mathews, and R.W. Leukel. 1953. A half century of wheat improvement in the United States. Adv. Agron. 5:1–151.

Slate, W.L. 1952. Agronomic research: Old and new. Agron. J. 44:1–3.

Swanson, A.F. 1958. Mark Alfred Carleton: The trail's end. Agron. J. 50:722–723.

ADDITIONAL READING

Swanson, A.F. 1958. Mark Alfred Carleton: The trail's end. Agron. J. 50:722–723.

Ball, C.R., and C.W. Warburton. 1925. Mark Alfred Carleton. J. Am. Soc. Agron. 17:514–516.

De Kruif, P. 1928. The wheat dreamer—Carleton. p. 3–30. *In* Hunger fighters. Harcourt, Brace & Co., New York.

Dies, E.J. 1949. Mark Carleton—Wheat explorer. p. 141–149. *In* Titans of the soil. Univ. of North Carolina Press, Chapel Hill.

Freeman, E.M. 1929. Mark Carleton 1866–1925. Phytopath. 19:321–325.

Isern, T.D. 2000. Wheat explorer the world over: Mark Carleton of Kansas. Kansas History 23:12–25.

Historical Accounts

C. R. Ball	 J. A. Bizzell	 H. C. Bolley	 B. E. Brown	 Lyman Carrier	 G. A. Crabb
 E. O. Fippin	 F. D. Gardner	 A. F. Gustafson	 J. N. Harper	 Alvin Kezer	 A. F. Kidder
 J. G. Lipman	 T. L. Lyon	 A. G. McCall	 M. F. Miller	 C. A. Mooers	 R. A. Moore
 M. L. Mosher	 Oswald Schreiner	 H. L. Shantz	 C. F. Shaw	 L. H. Smith	 W. H. Stevenson
 R. W. Thatcher	 J. D. Tinsley	 H. G. Wheeler	 A. R. Whitson	 A. T. Wiancko	 C. G. Williams

E. L. Worthen

History of the Organization of the American Society of Agronomy

The history of the American Society of Agronomy traces to a date preceding its organization. There was a period of preparation. A time when conditions were so shaping themselves that the formation of the Society was a logical development. It was this that insured the immediate success of the organization.

During the last few years of the nineteenth century and the early part of the present one there began in the agricultural colleges a disintegration of the old departments of agriculture into units of more limited range. Among the offshoots was agronomy. The same movement took place in the Federal Department of Agriculture. Carleton in the first presidential address to this Society, stated that in 1900 there were only three agronomists in the agricultural colleges. Appointments in Agronomy increased so rapidly that in 1908 there were 99 persons holding that title. An equally rapid development of the subject took place in the U.S. Dept. of Agriculture. In 1900 the term was not used in the Federal Department but a few appointments of agronomists and assistant agronomists were made in 1901. By 1908 these appointments had increased to at least 100. I do not intend to convey the impression that agronomic work was not done before the closing years of the last century. Previous to that time, however, it was conducted in a somewhat restricted way by persons who bore other titles, and who in most instances conducted other lines of work as part of their official duties. The segregation of agronomy into a separate unit and the specialization which went with it created a need on the part of agronomists for contact with men in similar lines of work.

Many departments of agronomy in the agricultural colleges included such subjects as farm mechanics and farm management. However, these subjects were never considered to be within the field covered by this Society. Article II of the Constitution reads: "The object of the Society shall be the increase and dissemination of knowledge concerning soils and crops and the conditions affecting them." Accordingly, the papers presented at meetings of the Society did not include any on farm mechanics or farm management, even during the early life of the organization.

As a consequence of the advanced stage of development of agronomy at the time of the formation of the Society, the organization experienced no difficulty in maintaining its existence at any subsequent period. The 101 charter members became 121 by the close of the first year. In 1909 the roll was increased by 25 new members and in 1910 by 46 new members. In the latter year the total membership was 176, only 14 persons having fallen by the wayside during the first 3 years. The roll published in 1926 contained 653 names and today the membership comprises 949 persons. In addition to every state and territory in the Union, and nearly every province in Canada, the following countries are represented on its rolls: Argentine, Australia, Brazil, British West Indies, British Guiana, China, Ceylon, Colombia, Cechoslovakia, Cuba, Denmark, England, Estonia, Dominican Republic, Dutch East Indies, Egypt, Finland, Fiji, France, Germany, Greece, Holland, Honduras, Haiti, India, Ireland, Italy, Japan, Jugoslavia, Mauritius, Mesopotamia,

Reprinted in part from
J. Am. Soc. Agron. 25:1–9, 1933

Presented as the report of the Historian of the Society as part of the anniversary program commemorating the first twenty-five years of the Society's life.

T. Lyttleton Lyon
Head of the Department of Agronomy
Cornell University, Ithaca, New York

Thomas Lyttleton Lyon, Secretary, 1907–1909, as pictured with his biography in Proceedings of the American Society of Agronomy *2:11–12, 1911.*

Left: Charter members of the Society, a composite prepared from the original photos published with this article.

Section page: Plowing under alfalfa to improve the soil, early 1900s.

Mexico, Morocco, Norway, Peru, Poland, Portugal, Roumania, Spain, South Africa, Sweden, Switzerland, Turkey, Uruguay, Russia, Wales, West Indies, and Federated Malay States.

The total number of memberships outside of the United States and Canada is 74. This does not include Hawaii, Phillipine Islands, Porto Rico, or other outlying portions of the United States, which are also represented.

The first move towards the formation of a national Society, entirely covering but limited to the field of agronomy, took place in Washington, D.C. Action was taken at one of the meetings of the Agronomic Seminar of the U.S. Dept. of Agriculture in the fall of 1907. Even at this early date the Seminar was a very active one. A committee consisting of M.A. Carleton, W.J. Spillman, C.V. Piper, E.C. Chilcott, and A.D. Shamel was appointed to ascertain the opinions of agronomists throughout the various states, and also of certain college presidents, directors, and others, concerning the advisability of organizing such a Society. The following letter which was sent out has not been printed in the records of the Society and is therefore recorded here:

The letters received in reply to this communication are in the archives of this Society. It is not known how many of the circular letters were mailed, but 58 replies were received. It is quite noticeable that many of the administrative offices were inclined to question the desirability of adding another society to the list to which they might be expected to pay dues, although they did not express the feeling in those words. However, the agronomists were almost unanimous in the opinion that a society of agronomists should be organized.

As a result of the canvass another letter was sent out on December 12 announcing a meeting to be held in Chicago on December 31, 1907, during the week in which the American Association for the Advancement of Science was to meet in the same city. This letter had attached to it the names of 38 persons who were willing to sign the call for a meeting and to assist in the organization of a Society. The letter was printed in the Proceedings (Vol. 1, page 6) of the Society, will not be reproduced here.

Last year a list of charter members of the Society now on the rolls was printed in the report of the Historian. This needs some revision and the list is therefore inserted here in full, with present addresses.

C.R. Ball, University of California
J.A. Bizzell, Cornell University
H.L. Bolley, N. Dak. Agr. College
B.E. Brown, U. S. Dept. of Agriculture
Lyman Carrier, Coquille, Oregon
G.A. Crabb, University of Georgia
E.O. Fippin, McLean, Virginia
F.D. Gardner, Pennsylvania State College

A.F. Gustafson, Cornell University
J.N. Harper, Atlanta, Georgia
Alvin Kezer, State Agr. College, Colorado
A.F. Kidder, Galesburg, Ill.
J.G. Lipman, Rutgers University
T.L. Lyon, Cornell University
A.G. McCall, U. S. Dept. of Agriculture
M.F. Miller, University of Missouri
C.A. Mooers, University of Tennessee
R.A. Moore, University of Wisconsin

M.L. Mosher, University of Illinois
Oswald Schreiner, U.S. Dept. of Agriculture
H.L. Shantz, University of Arizona
C.P. Shaw, University of California
L.H. Smith, University of Illinois
W.H. Stevenson, Iowa State College
R.W. Thatcher, Massachusetts State College
J.D. Tinsley, Santa Fe R.R., Amarillo, Texas
L.R. Waldron, N. Dak. Agr. College
J. M. Westgate, University of Hawaii
H. J. Wheeler, Upper Montclair, N.J.
A.R. Whitson, University of Wisconsin
A.T. Wiancko, Purdue University
C.G. Williams, Ohio Agr. Exp. Station
E.L. Worthen, Cornell University

So far as possible, photographs of these charter members taken somewhere near the date of the founding of the Society have been collected, and are reproduced herewith.

Covering as wide a range of subjects as soils and field crops, including the breeding of these crops, it was inevitable that papers

---

UNITED STATES DEPARTMENT OF AGRICULTURE
Bureau of Plant Industry

GRAIN INVESTIGATIONS

Washington, D. C., Nov. 30, 1907

Dear Sir:

At the last meeting of the Agronomic Seminar of this Department the undersigned were appointed a committee to propose the formation of an American agronomic society. This committee is authorized by the Seminar to invite other men to serve with them on a final committee which shall call a meeting of all persons interested in agronomy at Chicago during the time of meeting of the American Association for the Advancement of Science this coming holiday season, for the purpose of organizing such a society. You are hereby respectfully invited to be a member of this final committee. As soon as acceptances of these invitations are received a call for the meeting will be at once issued.

There are a number of reasons for the existence of such a society several of which are so self-evident that it seems hardly necessary to discuss them in detail at this time. It is probably sufficient for the present to call attention to the fact that although the study of field crops and their relation to soil and climate is so important and has come to be considered in recent years of more and more importance, there is no society in this country at present giving attention to this subject, except the Corn Association recently organized, which is limited in its field of action to *one* of the many important field crops.

It is suggested that the committee issuing the call for this organization meeting have, as far as possible, every thing ready to facilitate a speedy organization, thereby leaving more time for the society itself to give attention to other matters besides business. In addition to the suggestive outline of the purposes and plans of the society, a draft of the constitution and by-laws should be presented as a basis of action for the society and it is probable that even a provisional program could be gotten together. In view of the short time now preceding the holiday season it is highly important that we receive an answer from you as soon as possible. Please give your opinion of the desirability of forming such an organization and make any suggestions concerning the subject that come to mind. Give titles and time required for any papers you can present.

Signed:
M. A. Carleton
W. J. Spillman
C. V. Piper
E. C. Chilcott
A. D. Shamel

The letter as reprinted in the 25th Anniversary article, J. Am. Soc. Agron. 25:2–3, 1933

should be classified according to subject matter. Accordingly, the first program committee appointed was instructed to provide for two sections, one for papers on soils and the other for papers on field crops. This differentiation of the program was prompted, or at least insured, by a move already on foot at the time of the formation of the Society to start an organization in the Mississippi Valley to deal with the subject of soils. The persons who had conceived the plan then joined the Agronomy Society with the idea of incorporating this embryo organization in the larger society.

These two sections have remained active during the first quarter century of the life of the Society. No others have been formed, although proposals to do so have been discussed several times. Perhaps a substitute for further differentiation of the program has been found in the custom of holding symposia in some closely restricted field. The first symposium was held about 1920, apparently at a special meeting in December of that year. The original idea was to have a review of the literature of some subject, bringing it up to date, and discussing all of its phases. This was particularly useful to busy men who did not have time to do this themselves, and especially to those who were in colleges not well provided with library facilities.

Gradually, however, the form of presentation changed. It became less a review of literature and more a discussion of the speakers' own work on the subject assigned. Naturally the leader of the symposium selected speakers on account of their own interest and accomplishments in the field of research chosen for the discussion. As there are nearly always to be found several persons working in closely related lines, a symposium really constitutes a temporary section of the Society devoted to the presentation of research in a particular field. In consequence the call for subdivisions of the Society into many sections has become less marked in recent years. By means of symposia it is possible to have a temporary section on any subject desired.

There is no one to whom the Society is more indebted, or perhaps so much indebted, as to those persons who have borne the burdens that fall upon the editors of its publications. Editorial duties call for wide knowledge, powers of discrimination, tact, and courage, besides almost daily attention to the details of the work. At the second annual meeting of the Society, held in Omaha, December 7 to 8, 1909, it was voted that an editing committee of five be appointed, the Secretary of the Society to be, *ex-officio,* secretary of the committee, and that the proceedings of the Society be published in as large an edition as was possible with the funds in hand. Carleton R. Ball, who had been elected Secretary of the Society, thus became editor. Under his editorship were published volumes 1 to 4 of the Proceedings of the Society. These contained the papers presented at the meetings from 1907 to 1912, inclusive.

In 1913, Dr. Ball began publication of the *Journal of the American Society of Agronomy,* provision having been made by the Society for publication of a journal to contain papers in addition to those presented at the meetings. The Journal was issued quarterly in 1913. Dr. Ball continued the editorship through 1914. That year five numbers of the Journal were published.

In January 1915, C.W. Warburton began his duties as Secretary and Editor. The Journal was increasing in size nearly every year and the papers were becoming more technical. This made the editorial work more exacting. To relieve the Editor, the position of Secretary was combined with that of Treasurer in 1917. Dr. Warburton was thus able to devote more time to the editorship, which duties he continued to perform until the close of 1921. Six numbers of the Journal were issued in 1915 and also in 1916, nine in 1917, eight in 1918, and nine in 1919. In 1920 it dropped back to seven numbers, and in 1921 and 1922 the practice of publishing eight numbers was resumed.

With the beginning of 1922, Roscoe W. Thatcher became Editor. This position he held until the close of 1928. Most of that time he had the help of James D. Luckett, who was named Assistant Editor in 1923. That year was the first in which 12 numbers were issued and this has been maintained to the present time. Since January 1929, the editorship of the Journal has been entirely in the hands of Mr. Luckett.

Up to the close of 1930 there have been 1,322 separate papers printed in the Journal, in addition to records of the transactions, reports of committees, book reviews, resolutions, and other miscellaneous matter.

The holding of meetings and the publication of papers have been the two most important functions of the Society. The former of these has fallen on many shoulders, but the latter has been carried throughout the past quarter century by the four men I have named. During all this time the Proceedings and the Journal have been ably edited. On this function of the Society we may look with profound pride.

It has always been rather difficult to finance the Journal which has been almost entirely dependent on membership dues for its support. The Society began life with the modest annual fee of $2.00 per member. This was continued until 1908 when it was rather gingerly raised to $2.50. There was always a fear that a more adequate fee would result in reduced membership.

In 1917 there were 652 members, which was the largest roll attained to that date. In 1918 the number dropped to 509 and continued to descend until 1920, when it reached 435. This was so evidently due to the influence of the war, and its aftermath, that any effect of the 50 cent rise in dues is entirely masked.

By 1922 the membership had risen to 643. The Journal was having a hard struggle. A courageous committee recommended that the annual dues be increased to $5.00. The Society endorsed this stand and the increase took effect in the beginning of 1923. That year there was a loss of 82 members. However, in 1924 the roll increased by 16 members and in 1925 the membership stood at a higher figure than before the change in fee took place. From that time there has been a larger membership each year up to the present time. The larger revenue has made possible a great improvement in the Journal, which has doubtless been reflected in the increased membership.

The close of the first quarter century in the life of the American Society of Agronomy finds that organization in the most successful period of its career. The membership is the largest in its history, the meetings are more largely attended, and the papers published in the Journal are more scholarly than ever before. While the outlook for the immediate future does not encourage the hope of a pronouncedly continued increase in membership, owing to the general retrenchment in expenditures by the agricultural, as well as other scientific institutions, yet there is every reason to believe that the quality of the work will at least continue on its present high plane.

History of American Society of Agronomy First Fifty Years—1907 to 1957

This history aims: (1) to present the vision put into action of men who devoted and are devoting their lives to the science of agronomy and (2) to bring together a record of the results of their efforts and aspirations as exemplified in the development of the American Society of Agronomy from its organization in 1907 to its fiftieth anniversary in 1957.

Much of the history was written by members who were participants in the activities and events which they recount. The value of the history is enhanced by their personal contact and insight of questions considered.

The historian is deeply indebted to each one who contributed, namely, M.F. Miller, J.D. Luckett, G.G. Pohlman, D.S. Metcalfe, W.H. Pierre, and Emil Truog, and to those who aided in initiating and publishing this record of our society through its first half century of service to agronomy.

Several excellent reports of historical information have been published previously in the Journal.[1]

H. H. Laude
Professor of Agronomy
Kansas State University
Agronomist
Kansas Agricultural Experiment Station

FOUNDING OF THE AMERICAN SOCIETY OF AGRONOMY

M. F. Miller[2]

The American Society of Agronomy was born on December 31, 1907. It has gone through a half century of development without serious complications. Its expansion has really been phenomenal. Fifty years ago, the roll of its charter members was 101. The number of active members on its rolls September 30, 1957, was 3,258. The great increase in the number of papers presented annually, and the growth in its various lines of activity will be given elsewhere in this anniversary report.

The Society was organized soon after the term *Agronomist* came into use by the U.S. Department of Agriculture and by the Land Grant Colleges. However, at that early period there had been organized in the Department of Agriculture an *Agronomic Seminar,* which was meeting regularly in 1907.

At one of the late Seminar meetings of that year, a discussion took place regarding the desirability and possibility of organizing a national society composed of men in the fields of soils and field crops. There was already a Mississippi Valley organization, dealing with soils, but the group thought this might be drawn into an organization covering both soils and crops with an Agronomic name. As a result of these Seminar discussions, five members were selected as a committee to explore the situation and see what might be done.

The committee appointed were stalwarts in the Bureau of Plant Industry of the Department of Agriculture at that time. They were M.A.

Charter members of ASA present at 50th Anniversary Meeting in Atlanta, November 1957. L. to r., (seated) C.A. Mooers, H.L. Shantz, M.F. Miller; (standing) Alvin Keyser, E.L. Worthen, Lyman Carrier, and M.L. Mosher.

Reprinted in part from
J.Am. Soc.Agron., 54:57-69, 1962
Presented as the report of the Historian to commemorate the fiftieth anniversary of the American Society of Agronomy in 1957.

[1] T.L. Lyon (J.Am. Soc.Agron. 23:1035 and 25:1–9), R.I. Throckmorton (J.A.S.A. 33:478–479 and 1135-1140); F.D. Keim (Agron. J. 5:651–654).

[2] Dean and Director Emeritus, College of Agriculture, University of Missouri, Columbia.

Carleton, W.J. Spillman, C.V. Piper, E.C. Chilcott, and A.D. Shamel. Every early agronomist remembers these men and all present-day agronomists know them by reputation. All of them had illustrious careers.

After due consideration the committee formulated a letter outlining the ideas it had in mind and sent it to all of the men in the country who were working in the field of agronomy.

In answer to this letter, 58 replies were received. Practically all the technical agronomists approved the idea of calling such a meeting. Some of the men who were associated particularly with the administrative aspects of agronomy were somewhat skeptical as to the undertaking. However, since the approval by the technical agronomists was almost unanimous, the committee decided to go ahead and issue the call for the meeting.

A letter was prepared calling the meeting in Chicago on December 31 as had been suggested. This letter had attached to it "the names of 43 persons who were willing to sign the call." This letter, along with the names of the signators follows:

"It is requested by the undersigned that all persons interested in agronomy join with them in a meeting during the coming holiday season at Chicago for the purpose of organizing an American Society of Agronomy. The meeting for organization will be held Tuesday morning, December 31, in the buildings of the Chicago University, the particular place of meeting to be announced later.

"Although this meeting is called for the same period during which the sessions of the American Association for the Advancement of Science will be held, in order to secure better attendance, it will be determined at such time whether the organization shall be a section of this Association, an affiliated society, or take some other form. A suggestive constitution and bylaws will be submitted in order that business may be hastened and as much time as possible given to the reading of papers even at this first meeting. Eight to ten papers have already been offered. Titles and abstracts or the papers themselves should be sent at the earliest date possible either to Alvin Keyser, University of Nebraska, Lincoln, Nebr., or to M.A. Carleton, U.S. Department of Agriculture, Washington, D.C."

C.A. Alvord	Wm. D. Hurd	R.H. Rolfs
Alfred Atkinson	Alvin Keyser	A.D. Shamel
E.D. Ball	B.W. Kilgore	J.H. Shepperd
Wm. P. Brooks	E.R. Loyd	C.D. Smith
C.P. Bull	T.L. Lyon	L.H. Smith
M.A. Carleton	M.F. Miller	A.M. Soule
E.C. Chilcott	E.G. Montgomery	W.J. Spillman
L.A. Clinton	R.A. Moore	R.W. Thatcher
W.R. Dodson	L.A. Moorhouse	J.D. Tinsley
J.F. Duggar	H.A. Morgan	G.H. True
G.H. Failyer	C.L. Newman	E.B. Voorhees
F.D. Gardner	W.H. Olin	H.J. Wheeler
J.N. Harper	C.V. Piper	J.A. Widtsoe
C.G. Hopkins	W.J. Quick	C.G. Williams
T.F. Hunt		

This organizational meeting was called to order in the Botany Building of the University of Chicago at nine o'clock on the morning of December 31, with 43 men present. M.A. Carleton was chosen as temporary chairman and T.L. Lyon as temporary secretary. An informal discussion followed which covered the desirability of forming an "Agronomic Society," its nature, functions, objective, and the character of its meetings. The idea met with general approval and a committee of five was selected to deal with the matter of a permanent organization, and to report back at a two o'clock session in the afternoon.

This organizational committee was made up of M.A. Carleton, Chairman, H.P. Armsby, C.G. Hopkins, C.V. Piper, and T.L. Lyon. The meeting then adjourned to meet in the afternoon to hear the report of this committee. (This certainly seems like very rapid action, but it is evident that the members of the original Seminar committee had already formulated most of the essential features of the organization and that these were in Carleton's hands.)

At the two-o'clock meeting the committee formally recommended the organization of "The American Society of Agronomy," the object of which was: "The increase and dissemination of knowledge concerning soils and crops and the conditions affecting them." Annual meetings were recommended at places designated by a vote of the Society or as decided upon by the Executive Committee. Charter members were specified as all those forming the Society and paying dues before July 1, 1908. Provision was made for the formation of local sections of the Society of ten or more persons, by any three members, with the approval of the Executive Committee. A constitution and bylaws were also submitted for approval.

This report of the committee was adopted with the provision that the committee be retained and authorized to submit, at the next annual meeting, such additions and changes to the constitution and bylaws as might be deemed advisable.

The permanent officers elected for the ensuing year were: President, M.A. Carleton; First Vice-president, C.P. Bull; Second Vice-president, J.F. Duggar; Secretary, T.L. Lyon; and Treasurer, E.G. Montgomery.

The chairman appointed a program committee for the next meeting, consisting of A.R. Whitson for Soils and C.G. Williams for Crops. The committee on organization was instructed to report at the next meeting on a method of electing officers.

A resolution was introduced instructing the Executive Committee to offer any possible assistance to the Society for the Promotion of Agricultural Science in bringing about the affiliation of scientific agricultural organizations, including the Society of Agronomy, into a National Association for the Advancement of Agricultural Science. This resolution was presented by C.G. Hopkins and unanimously adopted.[3] Ten papers were presented at this organizational meeting, some by title only. Certain amendments to the constitution and bylaws were to be considered at the second meeting of the Society at Ithaca, N.Y., on July 9–11, 1908. However, consideration of the matter at that meeting was deferred until the third meeting which was held in Washington, November 17–18, 1908. At that meeting the amendments were presented and approved.

At the Ithaca meeting a committee was appointed, with the president as chairman, to consider the advisability of publishing a journal and report at the third meeting in Washington. At this meeting the committee reported that it did not seem wise for the Society to establish an organ of publication at that time. However, at the following meeting at Omaha, Nebr., on December 7–8, 1909, the committee recommended that a journal be established and this was approved. As a result, the first volume of the *Proceedings of the American Society of Agronomy* was published in 1910 with five members on the editorial staff. The first volume contained the papers presented at and the minutes of the first four meetings for the years 1907, 1908, and 1909.

One matter in the organizaiton of the Society is not entirely clear. This has to do with the names of those who can be considered officially as charter members. The early ruling that these should include all mem-

---

[3] *The Society for the Promotion of Agricultural Science was an old and highly respected organization in its day. It was Hopkins' idea to combine agricultural societies into a larger, over-all organizaiton to be designated as a National Association for the Advancement of Agricultural Sciences. Nothing ever came of this resolution, however, as the Association of Land Grant Colleges had taken over as the all-important agricultural organization early in the present century.*

bers who paid their dues by July 1, 1908, made 101 men eligible to this distinction.

T.L. Lyon, who was historian in 1933 published the names of 33 men whom he stated were charter members on the rolls of the Society at that time. They are listed on page 3 of Volume 25 of the Journal, along with photographs of 31 of them.

PUBLICATIONS: THE FIRST FORTY YEARS

J.D. Luckett[3]

The history of the publications of the American Society of Agronomy is adequately set forth in the reports of the successive editors and in the pages of the publications themselves through the years. But with the advent of the fiftieth anniversary of the Society, it may not be amiss to review this aspect of the Society's activities even at the risk of repetition of existing records.

It might well be said that the *publications* of a scientific society are *the society*, for certainly they are the chief tangible evidence of its existence and its vigor and vitality. There are other evidences, of course, such as meetings and committee activities, but the publications of the Society set the standard by which it can be evaluated by others and surely are the determining factors in establishing and maintaining the prestige that the society enjoys in the scientific world.

The American Society of Agronomy may well view with pride its heritage in the field of publication. Conspicuous in the history of this Society are two factors. One is that the Society has never sought nor enjoyed any form of financial sponsorship from outside its own ranks. Whatever has been accomplished has been done by sheer determination and hard work of the founding fathers and their successors. Another factor that emerges from the past is that most of the advances made in the Society's publications have been achieved the hard way, often in the face of financial adversities that might well have daunted less dedicated and intrepid folk. But, "let's look at the record."

Publication first authorized in 1909—The need for a medium of publication of agronomic literature was recognized from the very beginning and was the subject of discussion at the early meetings. Actually, authorization to proceed was given at a meeting held in Omaha, Nebr., in December 1909. A publication committee was appointed, consisting of C.V. Piper, chairman, C.R. Ball, secretary, G.N. Coffey, G.H. Failyer, and L.H. Smith.

To Ball fell the responsibility of collecting and editing the papers that made up the first volume of Proceedings, a most fortunate circumstance for the Society, as time was to prove, for he set a pattern and a standard that prevail substantially to the present time.

Sixty-nine papers had been presented at meetings of the Society prior to the publication of the first volume of Proceedings. Thirty-nine of these were published in volume 1, which appeared in 1910. Volume 2 contains the first "contributed paper, plus 15 papers presented at the 1910 meeting. It also contains the first index which covers volumes 1 and 2.

Four volumes of Proceedings were published, with distribution limited to members "not in arrears for dues." In 1911 the Society authorized the publication of papers other than those presented at the annual meetings, including two doctoral theses. The first instructions to contributors to the Proceedings appeared in volume 3.

The Journal is born—But from the first the Society envisioned a journal type of publication that would appear at frequent intervals and would provide a medium for current agronomic contributions. First mention of a journal was made at a meeting in Ithaca, N.Y., July 1908, when a committee consisting of M.A. Carleton, chairman, H.J. Wheeler, and T.L. Lyon was named "to consider the advisability of attempting to publish a journal." This committee reported later that year that it did not seem wise to establish "an organ of publication at that time, as it was possible that the affiliated societies of the Association of Agricultural Colleges and Experiment Stations would provide such a medium."

But this medium did not materialize, and by 1910 Ball was saying, "The need of a suitable medium for the prompt publication of papers relating to American agronomy is becoming increasingly evident. The time is now ripe for our Society to consider founding a high-class journal which shall adequately meet this need."

In 1911, he said, "there is an increasing need for such a medium of agronomic expression in this country... The sum of $1,000 per annum should prove sufficient to cover the cost of publication and of the current expenses of the Society. Three hundred and fifty members and a subscription list of 150 copies would provide this sum, aside from any revenue derived from advertisements." The fact that there were only 167 paid-up memberships at the time (at $2.00 each) and no evidence of any subscriptions, although there doubtlessly were libraries subscribing to the Proceedings, and that advertising revenues were nonexistent, did not in the least daunt the Executive Committee which named C.V. Piper, T.L. Lyon, and C.R. Ball to study the matter further. This was in 1912.

The report of this committee was presented in November 1912, at the annual meeting in Atlanta, Georgia, and stated in part that, "The publication of an annual volume of Proceedings will be discontinued with this volume. Early in 1913 the Society will begin the publication of a journal, quarterly at first and more frequently when circumstances warrant." And so the Journal was born!

The formative years—In format and general make-up, the Journal as it appeared first in 1913 was substantially the same as it was in 1948 when the transition to the present size and internal arrangement occurred. The name of *Journal of the American Society of Agronomy* was made official and Ball was continued as editor-in-chief. An editorial board was also established which was to be perpetuated to the present time in one form or another. C.V. Piper was named associate editor for crops, with L.H. Smith, C.A. Zavitz, and R.W. Thatcher as assistant crops editors. T.L. Lyon was associate editor for soils with C.E. Thorne, L.E. Call, and J.G. Lipman as assistant soils editors. All of these men played important roles in the Society in addition to guiding the new Journal through its formative years.

"Brief articles," later to be known as "Notes," were solicited, and a section on Agronomic Affairs was established as a regular feature of the new publication. The first book review appeared in volume 6. The Journal appeared quarterly in 1913 and five parts were issued in 1914. At the close of that year, too, Ball stated that pressure of other work made it necessary for him "to decline further honors" from the Society and requested that some one else be named secretary and editor of publications.

C.W. Warburton was named to the post, beginning January 1915, and immediately recommended to the Executive Committee that they appoint an editor "to relieve the secretary of a part of his present duties," but it was not until 1917 that the positions of secretary and treasurer were combined and Warburton named editor in fact.

Beginning in 1915, an effort was made to issue the Journal on a bi-monthly basis, "so long as the supply of material is sufficient."

Free reprints of papers appearing in the Journal were first issued in 1916, a practice that was to succumb to financial difficulties with the advent of World War II and the depression which preceded it.

But by 1916 Warburton was saying, "The next step in advance should be toward monthly publication. Our financial condition does not yet justify that step, but a partial advance can be made by issuing nine or ten 48-page numbers, these to appear monthly except in June, July, and August, or except in July and August, as the case may be."

---

[3] *Editor, New York State Agricultural Experiment Station, Geneva, and Life member of ASA.*

Volume 9, published in 1917, actually contained nine numbers, but serious financial troubles were just ahead with the coming of World War I. The dues were still $2.00, but they were increased to $2.50 at the 1917 meeting. Many members were going into the armed services, however, and Editor Warburton was to report that "The year 1918 has been the most disastrous period in the history of the Society. In previous reports there has always been recorded substantial growth; this one shows a decided decrease in membership." He also noted increasing costs of publication of 35 to 40% by 1919.

Crucial decisions—Warburton was finding his official duties mounting with the war years and a rapidly expanding extension program and in 1921 asked to be relieved as editor of the Journal. R.W. Thatcher, then at the University of Minnesota, was chosen to succeed him and took over the editorial responsibilities in January 1922.

The Journal was running a deficit and membership was still off, but it was decided to work toward monthly publication and to raise dues sufficiently to finance the Journal. Possible advertising revenue was again explored. By the end of 1922 the dues had been raised to $5.00 and monthly publication was begun in 1923. A recommendation by Thatcher that all papers presented at any meeting of the Society become the property of the Society for publication unless released was approved and provided a backlog of material against which to draw for the proposed monthly publication schedule.

The publication of complete symposia in single issues of the Journal was also approved at this time, and several issues were duly published. As the Journal became more firmly established and with a steady flow of contributions coming into the editor's office, both of these provisions were rescinded, that is, the proprietorship of papers presented at meetings and the publication of symposia, in order to expedite publication of contributed material. From that time on papers presented at the meetings took their place with contributed papers in the publication schedule and were subjected to the same rigid screening by the Editorial Board. Provisions for publications of Monographs and *Advances in Agronomy* under the sponsorship of the Society largely met the need for symposia-type material.

In mid-1922 Thatcher moved to Geneva, N.Y., to become director of the New York State Agricultural Experiment Station. There he found the experiment station publications being handled by the J.B. Lyon Company of Albany and entered into a contract with that concern to publish the Journal. This arrangement continued until 1924, when a contract was drawn with the W.F. Humphrey Press of Geneva, N.Y., which continued to print the Journal until recent years. The Society has been fortunate in the choice of its printers. The New Era Printing Company of Lancaster, Pa., printed the first volume of Proceedings and all other Society publications through 1921.

The year 1923 was notable in another respect in that the Journal showed an income of $697.60 from the first advertising to appear in its pages. So encouraged were the officers of the Society that they approved Thatcher's recommendation that they employ an assistant editor and advertising manager "at not to exceed $500.00 a year." J.D. Luckett, editor of the New York State Agricultural Experiment Station, was appointed to that post.

The Journal grows—By 1926, Thatcher was able to state in his report that, "The past year has been, in my opinion, the most successful in the history of the Journal." More papers had been published on more pages and there had been more income from all sources, including advertising, than ever before. The Journal had attained its growth and was set in the way in which it was to go for the next 20 years when forward-looking members of the Society were again to chart bold ventures of publication and expansion of Society activities which were the forerunners of today's splendid mediums of agronomic publication, both technical and popular.

Thatcher resigned his position at Geneva in 1927 to become president of what was then the Massachusetts Agricultural College, at the same time relinquishing his post as editor of the Journal.

The next 20 years—The Society came upon its next editor largely by following the line of least resistance. The assistant editor and advertising manager had been assuming more and more of the routine operation of the Journal during the last two or three years of Thatcher's tenure as editor. Upon his resignation in 1927, therefore, the executive committee named J.D. Luckett as editor, effective in January 1928. This association was to continue for 20 years.

Matters started off well with the new editor. One of the first projects was the preparation of a cumulative author and subject index of the first 20 volumes of Proceedings and Journal which was published in 1928. It was followed by two 10-year cumulative indices published in 1938 and 1948, respectively. The 1928 volume of the Journal also set a new record in number of papers and number of pages published. And then the depression struck!

A committee consisting of C.W. Warburton as chairman, J.G. Lipman, and W.L. Slate was set up to study the publication and financial situation and its reports ran through three volumes of the Journal. The final report and recommendations made at the 1933 annual meeting brought about restrictions on length of papers with provision for surcharges for extra pages, charges for illustrations, and other changes.

The editor's report for that year reflects the situation. "It is with a sense of relief that we approach the close of 1933 with the Journal still intact," states the report. "We feel certain that at no time in the history of the Society has the outlook for the Journal been so distressing... What with closing banks and falling revenues, the Journal was faced early in the year with a financial stringency that has continued to the present time. We are happy to report, however, that 12 numbers of the Journal have been financed out of 1933 revenues and that we shall enter upon the new year without a deficit." And from that point on, the Society and the Journal have never been in serious financial difficulties.

Another significant action affecting the conduct of the Journal occurred at the New Orleans meeting in 1939 when recommendations from what was then known as the editorial advisory committee, consisting of Richard Bradfield, R.J. Garber, Merle T. Jenkins, and I.L. Baldwin, for the creation of an editorial board were approved. This board was to consist of an editor and two associate editors, one for crops and one for soils, the three to be appointed by the executive committee. The associate editors, in turn, were to select consulting editors who were recognized specialists in their fields. This board had complete jurisdiction over the screening of papers submitted to the Journal, with the editor responsible almost exclusively for the editing and actual production of the Journal. With the increasing complexity of the material being submitted to the Journal, it is our belief that this action did more than any other one thing to maintain confidence in the Journal and to ensure the high professional standards which characterized the Journal through the years.

During the first 40 years of the Society's life, a total of 3,280 papers have appeared in the Proceedings and the Journal. This is exclusive of numerous notes, book reviews, news items, announcements, indices, and other miscellaneous matters.

And so after the struggles of the early years, a major depression, and two world wars, the Society took yet another major step forward in expanding its horizons and enlarging its publication activities with the appointment in 1945 by President F.W. Parker of a committee on "Policy and Program," with W.H. Pierre as chairman. The recommendations of this committee and how they have been implemented are recounted elsewhere. The Journal, for the first time in its history, acquired a full-time editor and took on a more modern format. Even the name was changed to "Agronomy Journal," to bring it into line with present-day usage.

The Society and its publications have come a long way since 1907 and, if the future can be forecast with any degree of accuracy in the light of the past, the next 50 years will see the American Society of Agronomy forge ahead to even greater heights of service and accomplishment to American agriculture.

PUBLICATIONS: THE NEXT TEN YEARS

Publication activity of the Society was to expand and undergo many changes toward the end of the first half century.

The Journal continued to carry technical papers and to be the organ for recording the business of the Society. In the 10 years, 1948–1957, 1187 papers were published in the Journal and paid subscriptions rose to 3210 as of September 30, 1957.

The *Soil Science Society of America Proceedings*, which had previously been published as a single annual volume, began to appear in 1952 as a quarterly journal. Although in new form, the Proceedings retained its name and sequence of volume numbers. It continued to publish papers presented at the annual meetings of the Society and was open to other papers submitted by members. In 1957, the Proceedings publication schedule was increased to six issues a year. The management and editorial responsibility of this journal is in the Soil Science Society of America—the publication activities being located in the Central Office of the two societies in Madison, Wis.

A new magazine—The Society's committee on Policy and Program reported in 1946 that "it was generally agreed by the committee that a nontechnical publication was needed to reach the many workers interested in the applied phases of agronomy, including workers in the extension service, soil conservation, seed industry, fertilizer industry, and others whose primary interest is in the application of agronomy in the field." Later the same year that committee proposed that a mail vote of the members be taken after they had received a preview issue of the proposed new publication. The following year a committee of which K.S. Quisenberry was chairman, reported favorably on the new proposed publication provided the Society would employ a full-time "executive secretary and/or editor." Thus, the establishment of a central office with full-time staff and the expansion of publications moved forward hand in hand.

The new magazine, *What's New in Crops and Soils*, first appeared in October 1948, with nine contributed feature articles that conformed to the objectives of the magazine to present concisely authoritative material, well illustrated and interestingly written. *What's New in Crops and Soils* has been published nine times a year. In addition to the contributed feature articles, there have been many short items in each number, all dealing exclusively with crops and soils topics. It was the intent of the Society that *What's New in Crops and Soils* should aid in dissemination of knowledge concerning crops and soils and the conditions which affect them. That this purpose has been attained is indicated by the magazine's large circulation of 18,722 as of September 30, 1957.

Monographs and Advances—As agronomic research increased, it became increasingly difficult for many of those involved in one way or another in the theory or practice of soil management and crop production to keep themselves even reasonably well informed of the newer developments in any but their immediate fields of activity. Also students were finding it more difficult to examine and organize the numerous reports of research. A committee appointed in 1941 to study this question reported "there is a real need for a series of monographs covering certain important but specialized fields in field crops and soil science." The monographs committee after further study found that some subjects covered broad areas and required extensive treatment to properly present the available scientific knowledge, while others were of lesser scope and could be properly dealt with in shorter manuscripts. Therefore, two series of publications were arranged, the larger subjects to be published in a series of monographs and the less extensive topics grouped in annual volumes.

Agronomy Monographs, volumes 1 through 6, were edited by A.G. Norman, chairman of the monographs committee, and were published through the years 1949 to 1956 by Academic Press, Inc. The responsibility of publishing Agronomy Monographs beginning with volume 7 in 1957 was taken over by the Society.

The shorter reviews of subjects in agronomic science have been published under the editorship of Dr. Norman by Academic Press, Inc. under the title *Advances in Agronomy*. The objectives of Advances is to survey and review progress in agronomic research and practices. Quoting Norman, "The articles are written by specialists. They are critical and reasonably comprehensive in treatment. They are written primarily for fellow agronomists (including students) across the hall and across the continents who also find it difficult to keep well informed in all phases of agronomy." The 9 volumes of Advances published from 1949 to 1957 contained the reviews of 80 topics in agronomic science.

Agronomy News Letter—In 1956, the *Agronomy News Letter* was initiated. The board of directors recommended that at least four numbers be published and circulated to all active and associate members of the Society the first year. This publication, the board emphasized, is not to be Considered another periodical journal, but instead a "bulletin board" for members bringing "timely news and events concerning your work, your Society and your profession." It is now published six times a year.

DIVISIONS OF THE SOCIETY

G. G. Pohlman[4]

From the beginning of our Society, its leaders have had in mind the establishment of groups within the Society in order to provide greater participation by Agronomists in the affairs of the Society. The groups include (1) local sections, (2) regional branches, and (3) subject matter divisions.

Local sections—Article V of the constitution adopted December 31, 1907, is entitled "Local Sections" and provides that "any three members may, with the approval of the executive committee, organize a local section of ten or more persons, making its own rules for associate membership, providing its members shall become active or associate members of the Society." This was amended in 1911 to provide that local sections pay 50 cents per member to the American Society of Agronomy.

Early in 1913 there were 27 institutions which had the prerequisite three members to organize a local section. By the close of that year local sections had been organized at Cornell University, at the U. S. Department of Agriculture, Washington, D.C., and at the Kansas Agricultural College. In 1915 a local section was organized by the New England Agronomists and others were established at Iowa State College and at Ohio State University. The following year local sections were organized at the Georgia State College of Agriculture, South Dakota Agricultural College, and North Carolina State College.

During World War I the local sections became inactive, but following the war President Warburton urged their reactivation. The record in the Journal shows reactivation of the New England section in 1921 and a report from the Texas section in 1922. The Journal carried regular reports from the New England section until 1930 when it was reorganized to become a regional branch, and occasional reports of meetings of the Iowa section. Apparently all other local sections were inactive following 1922. Replies to a recent questionnaire from our executive secretary showed there are at present five local sections as follows (date of organization given in parentheses): Iowa (1915); Florida (1939); Georgia (1947);

---

[4] *Head, Department of Agronomy, West Virginia University, Morgantown.*

Hawaii (1951); and Mississippi (1953). The establishment of local sections is being considered in several other states.

Regional branches—Although the organization of regional divisions as a part of the Society was not provided for in the original constitution, the Society recognized the desirability of meetings which would serve members in the various parts of the country. Summer meetings at Cornell in 1908 and at Michigan Agricultural College in 1912 were, in a sense, regional meetings. These meetings were organized by the Society and held in conjunction with the Graduate School of Agriculture of the U.S.D.A. In addition, special or regional meetings were held in conjunction with the National Corn Exposition at Columbia, S.C., in 1913 and at Dallas, Texas, in 1914. By 1912 the need for additional regional meetings was recognized and five regions were suggested. These were: (1) The Cotton Belt States; (2) The Northeastern States; (3) The Upper Mississippi Valley; (4) The Great Plains; and (5) West of the Rockies. In making this proposal, program meetings were urged so that every member of the Society could attend one program meeting each year. The Great Plains section organized such a meeting to be held at North Platte, Nebr., August 19–21, 1913. This is noted in the Journal as the first sectional meeting and included 10 states. Reports of later meetings indicate that the Great Plains section became a part of the Great Plains Cooperative Association and that meetings were held in conjunction with this association until interrupted by World War I in 1917.

In 1914 the secretary was authorized to arrange a regional meeting with the New England Agronomists if a program meeting were assured. However, later reports list the New England section as a local section rather than a regional division of the Society.

A three-day conference of Western Agronomic Workers was held at Logan, Utah, in 1917 and arrangements were made for a meeting the following year. Meetings of this group continued until 1923. In the meantime a Western Society of Soil Management was formed in 1922. This group did not affiliate with the American Society of Agronomy.

The first meeting of the Western Canadian Society of Agronomy was held December 28–30, 1920, at the University of Alberta. Annual meetings were reported until 1924 when the group affiliated with the American Society of Agronomy.

The Association of Southern Agricultural Workers at their meeting in 1921 voted to organize the Southern Section of the American Society of Agronomy. When presented to the Society, it was noted that the Constitution did not provide for such affiliation and the matter was referred to the executive committee with the suggestion that the constitution be amended. Apparently this proposal set in motion the action which later led to the regular establishment of regional branches of the Society.

The constitution of the Society was revised in 1922 to provide for the establishment of geographical sections of the Society and representation on the executive committee from each geographical section of the Society. The geographical divisions were specified as North Atlantic, Southern, Corn Belt, and Western.

Actual organization of the geographical divisions as provided in the amended constitution began with a summer meeting of the Corn Belt Workers in Agronomy held at Urbana, Illinois, in 1923. The Corn Belt section was officially approved at the 1923 meeting of the American Society of Agronomy. In 1924 the Western Agronomic Workers became the Western Branch of the American Society of Agronomy. The name was changed to Western Society of Crop Science in 1954. At that time it was noted that this society will serve as the Western Branch of the American Society of Agronomy and that its officers will alternate with the officers of the Western Society of Soil Science in matters of business with the American Society of Agronomy. The Western Canadian Society of Agronomy also became affiliated with the American Society of Agronomy in 1924. The Agronomy section of the Association of Southern Agricultural Workers became officially affiliated with the Society in 1926. Expansion of the New England section to the Northeastern section occurred in 1931. The various regional branches have continued to be active except for the Western Canadian Society of Agronomy which appears to have disbanded.

The type of meetings held by the various regional groups has been variable. The early meetings were largely program meetings but summer field meetings became popular about the time of the organization of the present regional branches. The Corn Belt section and the Northeastern section emphasized field meetings, the Western section combined field and program meetings and the Southern section had two meetings, one a summer field meeting and the other a program meeting. The increased demand for program time at the national meeting led the Society to urge programs at regional meetings. As a result, programs were included in the meetings of the North Central Branch starting in 1955 and in the meeting of the Northeastern Branch in 1957.

The regional sections became regional branches with the revision of the constitution in 1952. The Corn Belt Branch was renamed the North Central Branch in 1954.

Subject matter divisions—Even though our Society was small and the interests of the members not as diversified as they are at present, the first program committee appointed consisted of A.R. Whitson for Soils and C.G. Williams for Crops. However, the early programs were not divided into soils and crops divisions, and it was not until 1919 that further reference is made to program committees with members from both crops and soils.

The first division of the program into soils and crops divisions appears to have been in 1924 when the program for the second day had separate sections listed for these groups. In 1925 an Extension Symposium was listed in addition to programs on crops and soils.

A special committee on Reorganization of the Society was appointed by President Miller in 1930. Among the various topics considered was the sectionizing of the Society. Their report on this is as follows:

"There has been a strong demand from various quarters for sectionizing the Society. It was early discovered, however, that the committee could not agree on more than two sections, a *Soils Section* and a *Crops Section*. This proposal was therefore submitted to the departments interested and of the 34 departments voting on this proposal, 28 were in favor of it and 6 were opposed. Several expressed no preference. The committee has agreed unanimously to this proposal and it wishes therefore to recommend that the Society be divided into two organic sections—a Soils Section and a Crops Section, and that each section be allowed to work out such subsections as it sees fit."

In response to this recommendation the members at the business meeting approved "that two special committees be appointed by the President to prepare the programs for the next annual meeting and formulate plans for the operation of the sections."

As a result of this report a new constitution was presented to the members at the annual meeting in November 1932. The proposed constitution was adopted with minor changes and a Crops Section and a Soils Section became officially recognized.

Following the committee report presented in 1931, special committees were appointed to proceed with the organization of the Soils and Crops Sections. At the November 1932 meeting the organizing committee for soils, consisting of S.A. Waksman, R.I. Throckmorton, M.F. Morgan, and M.F. Miller, chairman, presented a set of bylaws for governing the section. These provided for meetings, officers and subsections. The bylaws as adopted are a part of the committee report (24:1011). The first officers elected were Richard Bradfield, chairman, and M.F. Miller, secretary. During the next year, subsection programs were organized in Soil Physics and Chemistry, Soil Biology, and Soil Fertility.

The Crops Section organizing committee of M.A. McCall, George Stewart, and R.J. Garber, chairman, also made recommendations for the formulation of a Crops Section but no subsections were suggested. M.T. Jenkins was selected as the chairman to organize the program for 1933.

Soils Division—Inasmuch as there were now three soils organizations, the Soils Section of the American Society of Agronomy, the Soil Survey Association, and the International Society of Soil Science, a committee was appointed during the 1933 meetings to examine the bylaws and to recommend such changes as needed to provide a single association of soil scientists. This committee reported that they were unable to formulate any satisfactory plan or agree on any form of single association of soil scientists and accordingly no plan was presented. They recommended that a new Committee be appointed. This recommendation was approved and the new committee consisting of two members each of the American Soil Survey Association, the International Society of Soil Science, and the Soils Section of the American Society of Agronomy was appointed. This new committee submitted a statement and recommendations (J.A.S.A. 27:947–956) to be considered at the December 1935 meeting of the Society. Copies of this report were submitted by mail to all members of the American Society of Agronomy and the American Soil Survey Association. The replies indicated a large majority in favor of the formation of a single soils society. The report was discussed and approved at a joint meeting of all interested in soil science at the meeting on December 4 and later approved by both the Soils Section of the American Society of Agronomy and the American Soil Survey Association. A joint committee was appointed to prepare a constitution to be submitted to members by mail ballot.

Without a constitution and bylaws, plans for the 1936 meeting were made jointly by the Soils Section of the American Society of Agronomy and the American Soil Survey Association. These included six sections as follows: Section I—Soil Physics; Section II—Soil Chemistry; Section III—Soil Microbiology; Section IV—Soil Fertility; Section V—Soil Genesis, Morphology and Cartography; and Section VI—Soil Science Applied to Land Use.

The new constitution was adopted at the meeting in Washington on November 17–20, 1936, and the Soils Section of the American Society of Agronomy and the American Soil Survey Association became the Soil Science Society of America.

The sections were continued as listed for the meeting except that Section VI was officially recognized as Soil Technology. Richard Bradfield was elected president and A.M. O'Neal, secretary. In keeping with the renaming provided for by the constitution, the sections were designated as divisions in 1947. These divisions have continued with minor revision. In 1949, Division IVA—Organic Soils and Division VA—Forest Soils were organized, Division IV was renamed Soil Fertility, Fertilizers, and Plant Nutrients in 1952 and a subdivision on Plant Nutrients was added in 1954. Division VI was later changed to Soil Conservation. With the adoption of the constitution in 1952, it became Soil Conservation, Irrigation, Drainage and Tillage.

Crops Division—Following its organization in 1931, the Crops Section prepared programs but did not attempt to separate into various subsections for several years.

In 1937 a proposal was made to organize the Crops Section into three subsections. After careful consideration the division into subsections on (1) Genetics, Cytology and Breeding; (2) Physiology, Morphology and Ecology; and (3) Miscellaneous Topics was approved at the 1937 meeting. Increasing demand from members for programs led to a reorganization into five sections in 1946. These were (1) Breeding, Genetics and Cytology; (2) Physiology and Ecology; (3) Production and Management; (4) Seed Production and Technology; and (5) Special Topics. In preparing the program for 1947, Section 5 was divided in two sections, (5) Turf and (6) Weed Control.

In 1949 a committee was appointed to study the desirability of organizing the Crop Science Division into a Crop Science Society. After a thorough study this committee made the following recommendations:

1. That the Crops Science Division be redesignated as the Crop Science Divisions with a president and vice-president, each to hold office for one year with the vice-president succeeding the president.

2. That the present sections be designated as Divisions of the American Society of Agronomy as follows: Breeding, Genetics and Cytology; Physiology and Ecology; Crop Production and Management; Weeds; Turf Management; Seed Production; and Technology.

These recommendations were approved unanimously.

A committee was appointed in 1953 to prepare appropriate bylaws for the Crop Science Divisions. The bylaws were approved at the 1954 meeting at St. Paul, Minn. These provided for the same divisions previously approved with only the renaming of the Division of Weeds to the Division on Weeds and Weed Control. Following the approval of the bylaws, an amendment was proposed to change the name from Crops Science Divisions to Crop Science Society of America. This change was approved in 1955 and the Crops Science Divisions became the Crop Science Society of America.

Agronomic Education Division—From time to time special programs were arranged for extension workers. In 1936 a committee of extension workers appointed to study the possibility of a regular program at the meetings reported that the extension workers polled had indicated a preference for participating in the programs on soils and crops.

Members interested in resident instruction likewise organized some programs but these were included in either the soils or crops divisions. Meanwhile, increased emphasis was being placed in undergraduate student participation in Society affairs. When the reorganization of the Society was approved in 1947, a provision was made for the formation of new divisions. The Agronomic Education Division was established as a tentative division. This became a permanent division in 1949 with three sections on Resident Teaching, Extension [Teaching], and Student Activities.

Other proposed divisions and groups—Other special groups within the Society have developed programs at various times. Divisions on (1) Agronomic Applications and (2) Plant Nutrients were requested in 1947 and tentative divisions established. These had regular programs through 1951. Because the subject matter in these tentative divisions was also covered in the divisions in several of the soils and crops subject matter divisions, the Agronomic Application Division recommended in 1951 that the applied phases of agronomy be included in sections of divisions already established. This action was approved and the tentative division on Agronomic Application was discontinued.

The Plant Nutrients Division was also discontinued in 1952 when the Soil Science Division IV was renamed and a subdivision on plant nutrients added.

A special program on Military Land Use and Management was arranged in 1955 and continued in 1956. This is not yet classed as a new or temporary division. The program arrangements are made directly with the vice-president of the Society in his capacity as program chairman.

---

[5] *Assistant Director of Student Affairs, Iowa State College (now Director of Resident Instruction. College of Agriculture, University of Arizona).*

STUDENT ACTIVITIES SPONSORED BY THE SOCIETY

Darrel S. Metcalfe[5]

In 1921, at the meeting of the American Society of Agronomy a committee was appointed to consider the advisability of promoting a national organization for students in agronomy.

At the annual meeting in 1923, the committee consisting of W.C. Etheridge, John H. Parker, W.W. Burr, W.L. Burlison, J.F. Cox, and J.B. Wentz, chairman, submitted the following recommendations:

1. That the American Society of Agronomy go on record as favoring a national organization of students in Agronomy, but as believing that the initiative in the formation of such organization should be left to the students.

2. That the Society pledge its whole-hearted cooperation and support in case an organization is started by the students with the view of finally bringing about some affiliation of a students' organization with the Society.

3. That the president of the American Society of Agronomy be given authority to appoint a committee of three to act in an advisory capacity with the students in case the students attempt to launch a national organization.

Society records do not indicate any attempt to develop a national student organization for some years following those recommendations.

National Student Section—Interest was revived later and in the spring of 1932 P.E. Brown, president of the Society, announced the appointment of a special committee to study the organization of a National Student Section of the American Society of Agronomy. Committee members were: H.K. Wilson, F.D. Keim, J.W. Zahnley, G.H. Dungan, and E.R. Henson, chairman (J.A.S.A. 24:506, 1932).

This committee, after having obtained the reaction of students at Minnesota, Nebraska, Illinois, and Iowa to the proposed organization, reported that, "Students in agronomy and related fields have very favorably received the suggestion of the National Student Organization to be affiliated with the American Society of Agronomy as a Student Section. Students believe that membership in such a national organization will stimulate their interest in the field of agronomy. A circular letter recently mailed to agronomists has brought 12 favorable replies with none against the proposed organization."

The committee approved the form of the organization used at Iowa State College as a temporary organization but felt that the definite plan of the student organization should be left to the student section with the assistance of a committee from the American Society of Agronomy. The committee recommended that the Society authorize the organization of a student section of the Society to be composed of students in Farm Crops and Soils and in related fields. The committee further recommended that a committee be appointed from the Society to proceed with the organization and direction of a student section of the Society.

Those recommendations were approved in the annual meeting of the Society in 1932 and the former committee was requested to proceed to carry them out. By 1941, charters in the Student Activities Section had been granted to 23 institutions.

During World War II, from 1942 to 1946, reduced enrollment in agricultural colleges made student organizations impractical. Therefore, the National Organization of Agronomy students became inactive, in fact extinct. It was not until 1947 that steps were taken to revive the student groups. The committee of the American Society of Agronomy almost literally "drew student names out of a hat" to set up a slate of national officers. Seniors were chosen who would not be eligible for office the following year and they were asked to meet in Chicago with delegates from agronomy clubs over the country during the Thanksgiving vacation. By 1953, there were 39 chapters with 1400 members. As of August 1957, there are 45 chapters.

In 1949 the Student Activities Section asked that it be made the third section in the Education Division of the Society. This was approved by the Society.

The bylaws adopted by the American Society of Agronomy in 1952 provided that student chapters may be authorized in colleges and universities that provide a 4-year curriculum in agriculture.

The Student Activities Section adopted a national constitution with bylaws in 1953.

Previous to 1950 the Student Activities Section held its annual meetings in Chicago at the time of the International Grain and Hay Show. Delegates to those meetings usually were members of the judging teams. There was little to those annual meetings besides business and the election of officers for the next year. At the 1949 meeting, after considerable discussion as to the pros and cons of meeting with the parent Society, it was "voted to hold a joint meeting this year with the parent Society." (Agron. J. 42:627–628. 1950). That practice has continued.

Members of the American Society of Agronomy who have served as chairmen of the student section are E.R. Hensen (1932–34), H.K. Wilson (1935–45), G.H. Dungan (1946–48), G.C. Klingman (1949), D.S. Metcalfe (1950–53), E.T. York (1954), R.M. Swenson (1955), H.B. Foth (1956–57).

Student programs with the parent Society usually consist of an opening session including a welcome from the president of the Society and reports from officers and committee. A half-day program is developed with guest speakers on such subjects as "Foreign Agriculture, Agricultural Trends, Prospect for the Future, Job Opportunities, etc." Another half-day program is devoted to viewing "slides" showing the activities of each club. Students have been urged to attend sessions of the parent Society.

After each national meeting reports of the Student Activities Section are printed in *Agronomy Journal*, usually in the December number.

As early as 1933, the American Society of Agronomy sponsored a national essay contest for undergraduate students in agronomy. An annual contest was held each year until 1942. In 1947, the annual contest was again sponsored after the interruption due to the war.

At first, rules required a technical essay of approximately 3500 words on an assigned topic. The rules were modified in 1950 providing for a semitechnical type of essay, and contestants were free to select their own topics. Under the present rules the essays are to be written in a style similar to that of articles used in *What's New in Crops and Soils*. Two or more suitable photographs must accompany each essay. The best entries in each contest are published in the above magazine.

Announcements of each year's contests are sent to each Land Grant college along with a copy of the current rules. Details of the contest are available from the advisor of the local agronomy club. Essays are judged by members of the student essay committee of the American Society of Agronomy and a representative of *What's New in Crops and Soils*.

Through the years various awards have been made, as is explained in the next section of this history. The International Intercollegiate Crops Judging Contest was first held in 1923 in Chicago in connection with the International Hay and Grain Show. The first National Intercollegiate Crops Judging Contest was held in Kansas City in 1929. Except for the war years, 1942 to 1946, the contests have been annual affairs, the one in Kansas City being held the Tuesday before Thanksgiving and the Chicago contest on the Saturday following Thanksgiving.

The American Society of Agronomy also stimulated both local and national interest of students in agronomy clubs by offering a contest,

---

[6] *Head, Department of Agronomy, Iowa State College, Ames.*

initiated in 1952, to determine each year the best agronomy club in the United States.

In 1957, there were 44 agronomy student clubs in 38 states affiliated as chapters of the Student Section of the American Society of Agronomy. Most of these are called Agronomy Clubs, but some have taken such interesting names as: Field and Furrow Club, Klod and Kernel Klub, Chaff and Dust Club, and Clover Club.

Complete information on how to organize a new club is given in the Handbook of the Student Activities Section of the American Society of Agronomy.

AWARDS AND RECOGNITION BY THE SOCIETY

W. H. Pierre[6]

The awards made by the American Society of Agronomy during the first 50 years of its history fall into three main categories: (a) awards to members of the Society for superior achievement in research, (b) awards to undergraduate students in agronomy and to student agronomy clubs, and (c) special recognition awards.

The Nitrogen Research Awards—The first research awards made to members of the Society in recognition of superior achievement were the Nitrogen Research Awards made from 1928 to 1931 through the financial support of the Chilean Nitrate Educational Bureau. The awards were administered by the Society through its committee on Nitrogen Research Awards.

A total of $5000 was distributed among the recipients for each of the first three years and $1000 between the two recipients of the fourth year. The awards were made to individuals for outstanding nitrogen research in relation to economic crop production. In addition to the cash awards each recipient received an appropriate certificate....[7]

Chilean Nitrate Awards for Research on the Rarer Elements in Agriculture—In 1934, the Society accepted the administration of an annual award of $5000.00 by the Chilean Nitrate Educational Bureau. The purpose of the award was to stimulate research on the rarer elements in relation to economic crop production. The awards administered by a committee of six members of the Society were made to individuals for outstanding research on the presence of the rarer elements in plants and their role in crop production and plant nutrition....

The Stevenson Awards—The Stevenson Awards were established in 1947 through the generosity of W.H. Stevenson and his wife, Rosatha Scott Stevenson. The objective of the donors was to encourage and stimulate agronomic research, especially on the part of younger members of the Society. Awards were made each year, 1948 to 1953, inclusive, to two members of the Society who had done outstanding research, one in the field of farm crops and the other in the field of soils. The awards consisted of $500.00 each in cash and a suitable certificate of recognition....

Student Section Essay Contest and Awards—The committee on the Student Section of the American Society of Agronomy initiated a student essay contest in 1933 and the first awards were made at the fall meeting in 1933. Two awards were given, the first consisting of $15.00 in cash and a subscription to the Journal and the second, a $10.00 cash award plus a subscription to the Journal. In 1934, a third award was also granted which consisted of a subscription to the Journal. These three awards were continued in 1935 and 1936. In 1937 a grant of $200.00 from the Chicago Board of Trade made it possible to increase the number of awards to eight. The first three awards then consisted of expenses up to $50.00 to attend the International Hay and Grain Show and the American Society of Agronomy meetings. In addition, appropriate medals were given the first three winners and subscriptions to the Journal. The other five winners received cash awards of $25.00, $20.00, $15.00, $10.00, and $5.00, respectively.

The Chicago Board of Trade continued to support the program to the extent of $200.00 annually from 1938 to 1942, inclusive. The grants remained the same except that only seven prizes were offered, the last four grants being $20,00, $15.00, $10.00, and $5.00, respectively.

Because of the war the contest was discontinued in 1943. It was reinstituted in 1947 and has been continued to the present time. The awards in 1947 were made possible by the grant of $100.00 received from M.A. McCall and the Northwest Miller. They consisted of six awards, ranging from $30.00 to $12.50.

Since 1948 the contest has been supported by an annual grant of $200.00 received from the American Potash Institute. The awards for the 10 winners in 1948 ranged in value from $50.00 to $2.50.

Since 1949 the three top award winners received credit up to $50.00 toward expenses to attend the national meeting of the society, and since 1953, the fourth, fifth and sixth place winners received awards of $15.00 each. Moreover, since 1952 all 10 winners received subscriptions to *What's New in Crops and Soils*.

Abstracts of the three prize winning essays from 1935 to 1949 inclusive were published in the Journal. In 1950 the essays were changed from a strictly technical to a semitechnical style and *What's New in Crops and Soils* has awarded $25.00 for each essay published in that magazine.

A total of 159 students have been given recognition through this program. Many of those recognized in the early years are now prominent members of the Society.

National Achievement Awards to outstanding Student Agronomy Clubs—Through the efforts of the student activities committee and the financial support of the American Plant Food Council, a contest was initiated in 1952 to recognize the activities of the Student Agronomy Chapters of the American Society of Agronomy. The criteria developed by the committee for rating the clubs were as follows: (1) Individual club activities, (2) club membership, (3) participation in the Intercollegiate Crops Judging Contest and National Student Essay Contest, and (4) cooperation with the national treasurer and corresponding secretary of the Student Activities Section.

The Club designated by the committee as the "Outstanding Agronomy Club" for the past year was awarded a trophy and $100.00 for delegate expenses to the national meetings of the Society by the American Plant Food Council, which, since 1955, has provided the funds to sponsor the program.

This award has been won by clubs from Texas A&M (1952, 1957), University of Nebraska (1943), Iowa State College (1954), and University of Illinois (1955, 1956).

Fellows of the American Society of Agronomy—During the early years of the Society there was increasing interest in having some means of giving recognition to members for outstanding professional accomplishment in agronomy and for service in forwarding the objectives of the Society. At the annual meeting in 1923 a special committee, of which J.G. Lipman was chairman, "recommended that the executive committee be authorized to elect to fellowship members of the Society whose professional record would warrant such recognition." The recommendation was adopted and the constitution was changed in 1924 to include "Fellows" as one of the classes of members.

The secretary in his 1925 report noted that the executive committee "agreed that not more than 10 to 15 should be nominated in any one year and that they should be members of the Society of at least 10 years standing." For some years the men to receive this honor were selected by the executive committee. Later a committee on Fellows was appointed to receive nominations from Fellows of the Society and to recommend a list of 12 candidates for consideration by the executive committee.

---

[7] *Ellipses mark omissions. Award recipients originally listed in this section and following have been omitted for brevity. See the original article. A complete list of award winners is included on the accompanying CD.*

That number was subsequently increased. Since 1948 each nomination is prepared on a specified form and is sponsored by two Fellows. The nomination committee, consisting of the past president of the Society as chairman and two Fellows from each division of the Society, submits from those nominated a list of candidates from which the Board of Directors chooses those to be honored as Fellows.

Fellows have been chosen each year since 1925 except in 1944 when the annual meeting was not held and the executive committee requested that no Fellows be elected. The number of Fellows chosen in any year has varied from 3 to 30; a total of 279 from 1925 to 1957. That number is less than one-tenth as many as the present active membership....

Special recognition—In 1925, W.M. Jardine was elected honorary member of the American Society of Agronomy. Dr. Jardine who was then Secretary of Agriculture had long been an active member of the Society.

Two lifetime memberships in the Society have been awarded as tokens of "the Society's deep appreciation for loyalty and unselfish contribution to its welfare." J.D. Luckett received this award for his services as editor of the journal, 1928 to 1948, and G.G. Pohlman was so honored for his services as secretary-treasurer, 1938 to 1948.

New awards program—This program includes American Society of Agronomy awards in Soil Science, in Crop Science, and in Agronomic Education, and Senior Student Awards.

At the annual meeting of the Society held in Cincinnati, Ohio, in November 1956, the Society approved the establishment of an awards program and adopted a general policy and procedure for carrying on this program, proposed by the committee on awards.

In order to recognize outstanding achievements by members of the Society, three agronomy achievement awards are made annually. These awards are known as the Soil Science Award, the Crop Science Award and the Agronomic Education Award. They consist of a cash award of $200 and a suitably inscribed certificate.

The second phase of the new program recognizes outstanding senior students in agronomy curricula at the various agricultural colleges and encourages student participation in local student chapters of the Society. A recognition certificate is given annually by the American Society of Agronomy to a senior college student from each of the local Student Agronomy Chapters who has been nominated by the Agronomy staff (crops and soils) of his institution on the basis of scholastic accomplishment, leadership ability, and contribution to his local club.

In order to finance this new program, the Society established an awards fund to which individuals, firms, and associations are encouraged to contribute.

The first three awards to members of the Society were made at the 49th annual meeting in Atlanta, Georgia, in November 1957, and the first senior student awards were made in 1958....

THE ENDOWMENT FUND

Emil Truog[8]

At the annual meeting of the board of directors of the Agronomy Society in 1951 at Pennsylvania State College the name of the Development and Endowment Fund Committee was changed to Endowment Fund Committee. The former ad hoc committee, of which the writer was chairman, had been constituted primarily for the purpose of raising funds to establish a semipopular magazine subsequently called *What's New in Crops and Soils*.

This function having been successfully completed (approximately $38,000 was subscribed by Society members, industry, and agricultural organizations), it was suggested by the writer that a standing committee with the changed name should be continued for the purpose of raising an endowment fund which might be drawn upon in case of an emergency, and during normal times would be allowed to accumulate, except that annual income from capital might in time be used for whatever purpose seemed desirable, such as special awards for outstanding achievement or meritorious service in the field of agronomy. Accordingly, a committee of 9 to 14 members, appointed annually by the President of the Society, has functioned for the purpose indicated, beginning in 1952.

After considerable discussion by the officers of the Society it was decided to open the campaign for an endowment fund by offering sustaining memberships at an annual subscription rate of $100. Offerings in this connection have been made, not to individuals, but to manufacturers and dealers in fertilizers, agricultural chemicals and machinery, processors of agricultural products, seed growers and dealers, and agricultural organizations. A sustaining member receives all publications of the Society. Also, at present, each member has unlimited attendance privileges at the annual meetings of the Society, but only one person, designated by the member organization, has regular voting rights.

Since the inception of the program, the number of sustaining members has increased from 26 in 1952 to 76 in 1957, indicating that the program has been eminently successful. To date, the amount received from this type of membership totals approximately $32,000.[9] Interest or earnings from investment of this fund is approximately $2,000. Direct lump sum contributions to the fund are, of course, also desirable and acceptable. The Society should ever be grateful to all who so generously contribute to this fund, which will materially help to assure the continued aim and purposes of the Society, namely, the advancement of the science and practice of agriculture throughout the world. After a sizeable fund has accumulated, it may become desirable to use a portion of the fund for the purchase or erection of an office and storage building which would serve as a permanent center for the Society and greatly facilitate and stabilize its various functions and operations.

While it is not practicable here to name the people who have served on the endowment fund committee or list the sustaining members who have annually subscribed to this project (these are all recorded in the *Agronomy Journal* and in Annual Meeting Programs), the writer does wish to thank all who have cooperated in this undertaking and to pay special tribute to L.G. Monthey, executive secretary of the Society (from 1948 to 1961), for his untiring efforts in furthering the project.

CHANGES IN CONSTITUTION AND BYLAWS

The first constitution and bylaws of the American Society of Agronomy has been presented in the first part of this history in connection with the founding of the Society. As the Society grew and developed its activities, many changes were made leading finally to the Bylaws of 1952 under which the Society has continued to work to the end of its first half century.

Major changes in the Constitution and Bylaws are mentioned here with reference to where each change is published in the Journal.

Amendments in 1911 (3:26–27)
Amendments in 1924 (16:813 and 15:531–532)
New constitution in 1932 (24:839–841 and 1026–1027)
Amendments in 1937 (29:1062)
Amendments in 1943 (35:1031)
Amendments in 1946 (39:176–178)
New bylaws in 1952 (34:637–639)

---

[8] *Emeritus Professor of Soils, University of Wisconsin, Madison.*

[9] *In 1961 the total in this fund was over $63,000.*

Subdivisions of the Society and the Regional Branches have their own constitutions and bylaws which are coordinated with the Constitution and Bylaws of the Society.

INTERSOCIETY COOPERATION

From the founding of the American Society of Agronomy through its first half century there has been continuing interest in cooperating with groups, organizations, and societies in other fields of agricultural science. Those relations usually have been maintained by committees or representatives of the Society and have afforded opportunity for the Society to participate in the development and application of science and technology in many national and international aspects.

No attempt is made here to name all of the cooperative activities in which the Society engaged but a few are mentioned to illustrate the Society's interest and effort.

At the first meeting of the Society a resolution was adopted aiming toward the affiliation of several societies, including the American Society of Agronomy, to form a National Association for the Advancement of Agriculture. In 1911, H.J. Wheeler was elected representative of the American Society of Agronomy on the Council of the Affiliated Societies of Agricultural Science.

The Society has maintained close relationship with groups whose activities are specifically related to agronomy, such as the seed industry, the fertilizer industry, the International Crop Improvement Association, and the American Soil Survey Association. The most important connections of the Society have been the close and almost integral associations with the Soil Science Society and the Crop Science Society since they were organized in 1932 and 1955, respectively.

The American Society of Agronomy has held membership in some organizations and has had representation on numerous intersociety boards and committees, for example: Agricultural Research Institute of the National Academy of Sciences; Division of Biology and Agriculture, National Research Council; Scientific Man-Power Commission; International Commission for the Nomenclature of Cultivated Plants; International Congress of Soil Science; National Science Foundation; Scientific Agricultural Societies; American Institute of Biological Sciences; Food and Agriculture Organization of United Nations.

Through contacts and associations such as these, the American Society of Agronomy has been aided in its work and has contributed materially to advancing the science and technology of agriculture.

STATISTICAL INFORMATION

The list of Society members with addresses was published annually from 1908 to 1912 in the Proceedings and for each of the next six years in the December number of the Journal. A more recent list of members with addresses, grouped by states or countries, was published in the July 1953 number of the Journal (45:341–352).

Historical reports in the Journal prepared by R.I. Throckmorton (33:1135–1140) and F.D. Keim (45:651–654) contain statistical data that are not included in this fifty-year history. They listed in tabular form the officers from 1907 to 1953, including president, vice-president(s), secretary, treasurer, editor, chairman Crops Section (later Crop Science Divisions), and chairman Soil Sections (later president Soil Science Society of America). They also reported by years the number of Society members. Keim, furthermore, included a list of deaths of members by years from 1931 to 1953. T.L. Lyon (23:1035) listed 32 charter members who were on the membership rolls in 1931. He reported a revised list of charter members of the Society on the rolls in 1933 and published the pictures of 31 of the 33 charter members he named (25:1–9).

Statistical data for the last four years of the half century may be found in the December numbers of the Journal.

Development of the American Society of Agronomy, 1958–1977

D.C. Smith
University of Wisconsin, Madison

The present effort is to provide a reasonably comprehensive summary of the development of the American Society of Agronomy (ASA) for the years 1958 through 1977. This is a 20-year period completing a total of 77 years in which the Society has grown extensively and the total activity has been greatly broadened and elaborated.

The following principal references have come to the present writer's attention, concerning the history of ASA (listed chronologically).

Lyon, T.L. 1931. Notes on the history of the Society. J. Am. Soc. Agron. 23:1035.

Lyon, T.L. 1933. History of the organization of the American Society of Agronomy. J. Am. Soc. Agron. 25:l–9.

Throckmorton, R.I. 1941. History of the American Society of Agronomy. J. Am. Soc. Agron. 33:1135–1140.

Keim, F.D. 1953. History of the American Society of Agronomy. (1941–1953). Agron. J. 45:651–654.

Laude, H.H. 1962. History of the American Society of Agronomy. First 50 years—1907–1957. Agron. J. 54357–69.

Since the previous summary the operations of the Society have been recorded in the *Agronomy Journal* in a quite complete form and those interested in specific aspects of development are referred to this source. One large operational area to be generally minimized is that concerning budget and finances. This aspect has been covered very well on an annual basis with tabular and explanatory summaries in the *Agronomy Journal*. It was suggested in 1965 that a special section on history be put in the *Agronomy Journal* once a year, under the heading "The Record."

In 1971 it was recorded by the Historian that collection of historical items was to continue and that assembly of pictures of historical significance was being initiated. These suggestions were not followed as intended.

The functioning of the Society is not dependent only on the contributions of the elected officers and the Headquarters staff. In addition to their efforts, hundreds of members contribute. Since it is not possible to name all important workers, names are being omitted in most instances. Similarly, it is difficult to give credit to the real originator of new ideas and new programs and naming of people is minimized. However, the Society has had many outstanding workers.

The activities, procedures, and programs of ASA, Crop Science Society of America (CSSA), and Soil Science Society of America (SSSA), the three associated societies, are markedly interlocked. In reviewing the history of ASA, therefore, development of the other two societies is highly relevant and vice-versa. Consequently the reports and histories of the other societies are of interest in the complete record.

J.F. Lutz was asked to serve as the first historian of the Soil Science Society of America in 1970, and to develop a chronological report. A

D.C. Smith.

Reprinted in part from
Agron. J., 72:227–240, 1980

Report prepared by the Historian, 22 Mar. 1979.

preliminary draft was available in 1972 and a finished report was published that same year.

ORGANIZATION

Increased membership and proliferation of activities and programs of ASA and its associated societies have necessitated the development of a rather extensive organization pattern....[1] The Crop Science Division of ASA became the Crop Science Society of America officially in 1955. The Society is governed by a board of directors and an executive committee, the latter having the authority to act for the board between board meetings. The board of directors of ASA includes the presidents, presidents-elect and immediate past presidents of the three associated Societies; a representative from each A, C, and S division; a representative from each regional branch and the executive vice president-treasurer. The executive committee is composed of the presidents, president-elect, and immediate past presidents of ASA, CSSA, and SSSA.

The activities and functions of the societies have been divided into 10 major areas for ASA, within each of which committees function presently as indicated.

Activity	No. of committees
Organization	0
Nominations	11
Operations and Finance	12
Publications	29†
Awards	20
Professional Advancement	5
Collaboration	30
Meetings	5
Scientific Affairs	3
International Activities	1

† Includes editorial boards and some temporary committees.

In 1958 there were 20 standing committees and some subcommittees. In 1977 there were 73 A (ASA), 41 ACS (ASA, CSSA, and SSSA), 4 AC, and 2 SC committees operating in the ASA. Both CSSA and SSSA have their own committee structures as suited to their individual Societies and their association with ASA.

The coding system for organization and committees was initiated in 1965 and subsequently has been modified and expanded. Each of these areas has a number range for assignment of committees and the letters A, C, and S are used as prefixes to indicate the participation of the respective Societies. Within the system, temporary and subcommittees may be coded also.

ASA has six broad subject matter Divisions:

A-1 Resident Education   A-4 Extension Education
A-2 Land Use Management   A-5 Environmental Quality
A-3 Climatology   A-6 International Agronomy

There is a subdivision (A-1-a) of Division A-1, for student activities. The CSSA has six divisions and SSSA nine divisions. Subdivisions may be designated as needed. New divisions are made after a trial period and with the approval of the Society.

---

[1] *Ellipses here and in the following denote omissions. See the original article. References to the supplementary material published in the journal following the original article have been omitted for brevity.*

Presently, cooperating organizations with ASA are the Agronomic Science Foundation, the Agronomic Administrators Roundtable, and the American Registry of Professionals in Agronomy, Crops, and Soils (ARCPACS).

Though the organizational patterns of the three societies are somewhat parallel, each of the Societies is autonomous and independent in all respects. The CSSA and SSSA members are also ASA members and the respective officers and members join to determine ASA activities and policies.

Specific activities of the societies are not always on a triangular basis and each Society functions independently when appropriate. When two or more societies have a mutual interest, Society officers usually cooperate to provide agreement and support for proposed actions.

Since 1961 there has been much effort directed to coordination of activities of the three Societies while still allowing independent action as seems desirable. In 1959 serious consideration of the relations of the three Societies and the merits of separation vs. remaining together were underway. Discussion was especially relevant due to the need for establishment of a permanent Headquarters which would be adequate for the growing Societies.

While at times there has been some divergence of interests and professional differences of opinion and viewpoint, there has been a willingness of crops and soils members to work together. This is the principal contributing factor in the cohesion and success of ASA.

Executive Committee

Development of a manual or "Bluebook" for guidance of Society officers was begun in 1961. This has been modified as needed and defines the duties of officers and committees. Reviews and revisions of specific items are continuous. The Officers and Committees Code is very useful in identification of functions, organization, and projects. A training session for new officers is held at the annual meetings.

The Intersociety Coordinating Committee composed of the presidents of the three associated Societies and the executive vice president, serves to expedite intersociety cooperation. It can make motions and recommendations but cannot make policy or take actions. The executive committee may take action on behalf of the board unless specifically prohibited by the bylaws. However, since the board of directors is the ultimate governing body of the Society, it may nullify any action taken by the executive committee.

The increasing workloads for the non-salaried officers, editors, and others have presented problems as the Society has grown. More frequent meetings of the executive and other committees, e.g., budget, have become necessary and regular schedules for such meetings have been established, based upon urgency and member availability and convenience. Provision of substantial aid to the Headquarters office has also been necessary.

Bylaws and Constitution

Bylaw and constitution revision has been necessary periodically and almost continually as needs have changed. In 1961 the Bylaws and Organization and Policy Committees were combined. In that year changes were proposed to authorize mail ballot decisions by the board of directors. Other changes were made and bylaw revision was approved by the board in 1962. A proposal was made in 1968 to revise the bylaws to clarify nomination, election, and terms of service of officers and board members. In 1970 the bylaws were again revised and approved. In 1974 revised bylaws were submitted to the membership for consideration and were approved in that year. Bylaws have been published in 1952, 1964, 1971, and 1974.

Reports from the Headquarters office at the annual meeting of ASA have been required to conform to the bylaws of CSSA and SSSA but not those for ASA.

When necessary, the Articles of Incorporation have been revised. In 1963 a restatement of the articles was mailed to the membership for consideration and approval. Again in 1968 it was necessary to revise the articles to meet the requirements for a tax-free status consideration with the Internal Revenue Service. Divisional and other changes are made as desirable. The Climatology Division was tentatively approved in 1963 and finally approved in 1966. A proposal was made in 1977 to change the name of the Division of Climatology to that of "Climate–Plant–Soil Relations." This was not approved.

In 1964 it was proposed that Division A-1 be reorganized into two Divisions, i.e., Resident Education and Extension education, on a trial basis. This was done.

In 1970 a new International Division was proposed, and it was established provisionally in 1973. In that year also a Division of Environmental Quality was accepted on a provisional basis. Both of these new divisions became permanent by 1975. In this year the proposal was made that a Division of Pesticides be considered. This was refused in 1976. In 1973 a decision was made not to sponsor a Weed Science Division.

Numerous problems relating to Society operation must be met as they arise. Action was considered in 1967 for members of ASA to vote in only two divisions of the Society. This proposal was not approved, and members may vote in any divisions they choose. In 1959 the Organization and Policy Committee established the nature of the membership of the board of directors. In 1977 the matter of succession, should the incumbent president become unable to continue, was reviewed. It was determined that the past president should succeed though this would require a change of bylaws.

Subdivisions of the Society and regional branches have their own constitutions and bylaws and these are coordinated with those of the national organization.

Regional Branches

The four regional branches of ASA have long been in existence and interest and support have varied from region to region. Annual meetings of the branches have usually been held..... While the membership of ASA is not distributed uniformly over the United States attention has been given to holding the national meetings alternately in the different regions. There has been discussion of the relations of the branches to the national organization and the desirability of coordination and support.

The Western Regional Branch of the Society is composed of the Western Society of Crop Science (WSCS) and the Western Society of Soil Science (WSSS). Both organizations are independent of each other and since 1959 have met separately; the WSSS usually meets in conjunction with the Pacific Division of the American Association for the Advancement of Science. Due to their commonality of interests, however, the two organizations agreed, in 1973, to hold a joint meeting approximately every 3 years.

Increased national support for the branches has been sought in several respects. Society assistance has been given in the publication of programs and abstracts of papers of the regional meetings. In 1965 aid was sought for publication of branch symposia. In 1974 plans were introduced to regionalize the publication of *Crops and Soils Magazine* to better permit more local interest and support. Officers of the regional branches have met to confer on the relation of branches with the national Society and a considerable degree of coordination exists. The branches have representatives on the ASA board of directors and nominations are made for Society president-elect.

Budget

Budgetary statements have been prepared and presented annually in the *Agronomy Journal*. Budget committees...have spent considerable time in association with Society officers and the Headquarters office staff in consideration of all aspects of the budget, including the disbursement of income and conformity to Internal Revenue Service requirements for a tax-free status. Investment of surplus funds, utilization in building expansion, and establishment of pension reserves have been included also in the considerations of various committees. Sharing of costs in operations and projects of mutual interest has been established. Headquarters office expenses sharing was apportioned as follows in 1977.

Crops and Soils Magazine	14%
Agronomy Journal	14
SSSA Journal	15
Crop Science	12
Monographs	6
Non-Journal services	39

These allocations are reviewed and adjusted annually as appropriate.

Generally the numerical growth of the Societies the past 20 years has been accompanied with increased income from dues, earnings and gifts, and a general broadening of the activities of the Societies. While not all projects have produced net earnings, funds available for the numerous programs have been adequate.

The total annual expenses of the associated Societies have more than doubled the past 10 years. Reserves have included the amount of a year's operational budget for severe emergencies. A cash flow estimate has been made at 30% of the annual budget.

An Investment Committee of three persons to represent all three Societies was appointed in 1969 and the first investments based upon its recommendations were made that year.

It has been necessary for Society officers to become concerned with the tax status, especially as income has accelerated and reserves have been attained. A special tax exemption status committee was operative. The tax status and publication income were studied in 1968 and in 1960 there was a request for a new tax classification. In 1970 there was a City of Madison decision to require real estate and personal property taxes from ASA. However, in 1971 the Society was exempted from real estate and property taxes.

The problem of handling memorial and similar funds was considered in 1962. In 1973 the Borlaug Book Fund and the Publications Donations Project originated, though the former began in 1971. In 1975 the Agronomic Science Foundation allocated $1,000 for use in the Publications Donations Project. In 1977 it was decided to limit the project to accepting and delivering back issues of the Societies' journals.

MEMBERSHIP

Since 1947 the membership has increased an average of 20% during each 5 years and doubled during the period between 1957 and 1967. There has been especially rapid growth since 1960 when membership doubled in the next 15 years. However, the total numbers may involve different types of membership counts in different years. The Society "lost" a considerable number of members in 1954 when the Weed Society of America was organized, though its first annual meeting was not held until 1967. Foreign membership has grown also as in 1969 this comprised 16.4% of the total. Such members participate in elections and other programs.....

In terms of subscription to the principal journals the percentage of crop scientists taking the *Agronomy Journal* has been relatively constant at about 45%. The percentage of soil scientists subscribing to the journal were 54% in 1965, 52% in 1970, and 51% in 1977....

State chapters of the Society have been encouraged, and that in Florida began in 1940. (It had its 26th annual meeting in 1966). There are eight state chapters at present. Existence of state chapters is thought to stimulate ASA membership and to support interest in the agronomic field. In 1974 it was voted to give certificates of membership to state chapters of ASA.

Recently Field Research Station Supervisors have become interested in Society meetings and have planned sessions in relation to the Extension Education Division.

Twelve classes of membership in ASA are recognized. These are

Active	Affiliate
Emeritus	Sustaining
Life (two categories)	Subscriber
Graduate	Fellow
Undergraduate (two categories)	Honorary
Associate	Honorary Life

Conditions for these memberships are delineated in the most recent Bylaws (Agron. J. 67:169–173, 1975). Members may be included in more than one classification. Sustaining memberships were established in 1952 when there were 26 members, each contributing $100 to the endowment fund. In 1957 there were 76 such members and in 1977 there were 139.

Each sustaining membership, whether an individual or a company, has one identified member representative with full membership privileges.

Student memberships were first made available in 1964 for graduate students. Such memberships were limited to 4 years. In 1977 there were 1,465 such members. Undergraduate membership is reviewed under student activities.

Life and Honorary memberships are awarded at the discretion of the board of directors and primarily for extensive service to the Society... Such awards have been infrequent and irregular.... W.M. Jardine was made an Honorary Member of ASA in 1925 and was the only one to 1941. In 1977 action was approved to elect no more than two Honorary members annually. Emeritus membership was recognized in 1958 and 105 were so identified. These included two charter members and six with 50 years or more of Society activity. An additional 35 members had served 40 years or more. Requirements included 25 years of active service, membership within the previous 10 years and age 65 or older.

In 1973 it was reaffirmed that a membership option may be held singly or in combination of CSSA, SSSA, and ASA. By virtue of the bylaws of ASA, membership is automatically granted to those holding membership in CSSA and SSSA. This membership occurs without dues assessment for ASA. This was first proposed in 1961.

AWARDS

The growth, increased financial capability, and outside support of ASA has resulted in a substantial proliferation of the award program in recent years. In 1957 the Agronomic Education Achievement Award was given. In 1959 the Organization and Policy Committee established Agronomic Service Award criteria and in 1961 the first award was given. Crop and Soil Science awards were initiated in 1957. In 1973 these awards were relinquished to CSSA and SSSA, respectively, and are now presented by these Societies. During the period 1963–1966 the award program was discussed extensively and the basis for awards reviewed. Prerogatives of the individual Societies were recognized in making awards and certain awards are controlled by CSSA and SSSA independently.

In 1967 action was begun to establish an International Agronomy Award and this was first given in 1968. The CIBA-Geigy Award was first given in 1970. The Edward W. Browning Award for contribution to improvement of food sources began in 1971, with ASA selecting the recipient and the New York Community Trust administering the award. An award for Agronomic Research accomplishment was established in 1975 and the *Crops and Soils Magazine* Award program began in 1963. In 1977 the Agronomic Education Award was divided into two awards, namely, the Agronomic Resident Education Award and the Agronomic Extension Education Award.

Action was taken in 1972 to base the awards of the Societies on the past 5 years except for the Edward W. Browning Award, which was to depend on the past 10 years. Individual awards receive separate committee consideration and the executive committee reviews and ratifies the selections made by the committees. Student awards will be considered under student programs.

Fellows

Fellows in the American Society of Agronomy have been sponsored since 1923, and the first Fellows were named in 1925. Including the year 1977, a total of 739 had been named, these about half being each for crops and soils.

In 1956 the board of directors authorized the executive committee to approve Fellows nominations. The committee charged with selecting Fellows has long been representative of both CSSA and SSSA and has as chairman the past president of ASA. In determining the final list of nominees to the board or to the executive committee all Fellows committee members vote on all nominees whether their interest is crops or soils. Beginning in 1976 SSSA initiated its own Fellows program. The CSSA has not had such a program. In 1975 announcement of award recipients from all three societies and Fellows of ASA and SSSA were combined into one brochure which was distributed at the time of registration for the annual meeting. This was a departure from the tradition of not announcing award recipients and Fellows until the presentation ceremony.

Definition of eligibility of members for Fellow nomination has been a continuing study. In 1969 a total of 36 were elected on the basis of 0.5% of the membership being eligible as established in 1961. In 1970 this percentage was reduced to 0.03. In recent years criteria modifications have been made as in 1959, 1961, 1972, and 1975. The following considerations are those of the latter year.

I.	Personal achievements and recognition (Items 1–4)	10 Points
II.	Professional contributions (Items 5, 6)	40
III.	Improvement of Agronomic Programs Practices, Products, and Other Service (Items 7, 8, 10)	30
IV.	Service to ASA, CSSA & SSSA (Item 9)	20
		100 Points

The Fellows committee disapproved the use of the concordance ranking method for determining the final ranking of the nominees and adopted the paired rank method for determining the final ranking of the nominees. The committee approved the use of time of service as a minor criterion for evaluation of the nominees.

Alphabetical lists of Fellows have been provided in successive ASA annual meeting programs. Numbers of Fellows per year have varied from 3 to 36 throughout the history of the program. Since 1957 from 10 to 36 have been elected annually with 10 in 1958 and 36 in 1969 from approximately 100 nominations each year.

Student Activities

In 1958 additional attention was being directed to promoting agronomy favorably to college students and the relation of educational background to performance in beginning courses in agronomy. Reports of committees on training agronomists and including recruitment, curriculum, faculty, and facilities and employment became available.

Consideration was given to the promotion of an organization pattern for undergraduate students in agronomy as early as 1921. By 1941 charters for student activity sections were granted to 23 institutions. These were discontinued during the period 1942–1946 and undertaken again in 1947. In 1957 there were 45 chapters with over 1,500 members. In 1977 there were 69 such clubs. Activities within the clubs have been diverse and have been guided and stimulated by local agronomists. Speech, essay, photography, and crops and soils judging contests have been included. These have also been recognized at the national level and suitable recognition for both individuals and teams has been provided by ASA, the Student Activities Subdivision, and private and corporate donors. Students have been encouraged to attend the national meetings of ASA. In 1969 national competition for an agronomy senior award was attempted and outstanding seniors in agronomy were identified in 44 states.

Illustrative material for use in encouraging agronomy as a major study has been provided in various forms. A *Careers in Agronomy* film was proposed in 1961 but not undertaken. Brochures on *Careers in Agronomy* were prepared and revised successively in 1961, 1963, 1967, and 1973. A portable traveling exhibit on career promotion was suggested in 1963, assembled and made available in 1966. Use of this exhibit was discontinued in 1977 after a considerable travel. Various efforts have been made to provide slide sets for student use and in 1969 a career promotion set was approved.

The Society has participated in supporting the Future Farmers of America (FFA) convention and career show, beginning in 1965 and also has aided in the program of the International Science and Engineering Fair in providing awards. The ASA was given a plaque in 1975 for its 10 years of participation and support of the Agricultural Career Show and the National Convention of FFA at Kansas City.

Several committees have operated both consecutively and simultaneously to foster planning and support for undergraduate students. High school visitation for career promotion and advising has been attempted. Descriptive information has also been provided to high school teachers to assist in advising and teaching. A file of three volumes of test questions for soils students' reference use was completed in 1977 and a parallel set of crops questions was in preparation.

Awards

The first Intercollegiate Crops Contest was held in 1923. Society support of the contest began much later. The National Essay Contest was begun in 1933, dropped in 1942, and reinstated in 1947.

In 1961 the American Seed Trade Association provided $200 toward Society awards and this was used for the student speech contest. The National Soils Judging Contest was begun in 1961 and in 1964 presentation of awards for both soils and crop judging was considered. A traveling trophy and plaques were approved for the national soils judging contest in 1966. Certificate awards to members of the first three teams at Chicago and at Kansas City livestock shows (Crops contests) were made in 1969. In this year also the National Student Agronomic Photography Contest was approved.

The National Senior Recognition program was established in 1958 and in 1969 a total of 31 schools participated. There were 117 participants in the Student Essay Contest.

The Regional Branches have become interested in recognizing achievements of undergraduate students and in 1977 the Northeastern branch gave award recognition to the best senior students. Support of regional crops judging contests has been proposed. An effort has been made to increase the visibility of student awards.

HEADQUARTERS STAFF

Since 1961 increasing work and responsibilities have required additions to the Headquarters staff. L.G. Monthey was Executive Secretary from 1948 to 1960 and established the Headquarters office in Madison. There were three employees in 1948, and C.J. Lewandowski and R.C. Dinauer arrived in 1953 and 1956, respectively. Matthias Stelly began as Executive Secretary 1 April 1961 and was made Executive Vice-President in 1970. The duties of the executive secretary were first defined in 1961 and were to be stated in the bylaws of each of the three Societies. While appreciation has been continuous generally, the board of directors passed a resolution in 1965 commending Dr. Stelly and associates in the Headquarters office for "good business management, administrative efficiency, and devotion to the societies they serve."

In 1963 a Committee on Public Responsibility and Information was established with workers primarily in the Washington D.C. area. This group was to provide information to the Societies, from the Capitol, as seemed helpful to the Societies' operations. This committee recommended establishment of an ASA Public Service Committee with a part-time secretary in Washington D.C. This was approved in 1965 and a part-time secretary, responsible to the committee, was established in 1967, in the person of F.P. Cullinan. This position was discontinued in 1969.

Consideration for provision of an assistant to the executive secretary began in 1966 and in 1969 such a person was found and appointed. He remained only briefly and D.M. Kral, previously assistant editor of *Crops and Soils Magazine*, succeeded him. Numerous other staff additions and changes have occurred.

A group insurance and pension plan was put into effect in 1956. However, this plan required a 3-year wait until the staff member became eligible. In 1961, however, life insurance (individual) was made available for the 3-year interim period. In 1974, the interim period was reduced to 90 days for eligibility for group term insurance.....

ASA Headquarters

It was considered to be urgent by 1958 that a permanent Headquarters building be provided for the working needs of the growing Societies and a committee was appointed to explore this matter. Madison was approved as a permanent Headquarters site in 1959 and membership approval for a new building was given in 1960. A site was chosen 1 April 1961, plans progressed and ground was broken in November, 1961. The site was 1.59 acres 94 × 360 × 418 ft., located on the southwest side of Madison, near a shopping center. The building was dedicated 29 Oct. 1962, though it was completed in July.

In 1965 the Budget and Finance Committee recommended the expansion of the building. Drawings were prepared and plans approved in 1966. Work on the new addition began in 1967 and was completed then. The total space provided was then 9,116 sq. ft. with the new addition being 3,276 sq. ft.

The matter of still additional space arose again in 1974 and a space increase of 25 to 50% was suggested. In 1975 an increase of 6,200 sq. ft. additional was proposed and this was approved by the board of directors. Bids for the expansion were opened in early 1976 and the new space was occupied on 13 October of that year.

PUBLICATIONS

The principal publication of the American Society of Agronomy until 1949 was the *Journal of the American Society of Agronomy*. In 1949 it became the *Agronomy Journal*, and the format was enlarged.

Beginning about 1957 the need for a journal to implement the handling of crop science papers was recognized. The matter was related to the individual and cooperative plans of all three Societies and was associated with the questions of permanent Headquarters establishment and housing for all three. Soils and Crops Sections of the ASA were identified in 1924 and a Soils Section was formally established in 1931. In 1936 its constitution and bylaws were ratified. The journal, *Soil Science Society of America Proceedings*, was begun in 1936 and by 1960 the SSSA had considerable assets. Meanwhile the Crop Science Division had continued as such and had not built up comparable funds. Publication of *What's New in Crops and Soils* (*Crops and Soils* since 1958) had begun in 1948 and was not self-sustaining. Finally in 1959 approval was given by the membership for the publication of a new *Crop Science* journal (begun in 1961) and the problems of adjustment with the new and older publications were resolved....

Agronomy News began in 1956 as a bimonthly publication and became a standard carrier for current communications.

Crops and Soils Magazine

This publication began in 1948 as *What's New in Crops and Soils*. In 1957 it had 18,722 subscribers, and in 1977 it had 23,114. Since its initiation it has been necessary for the Society to provide additional support beyond the publication's earnings. Promotion campaigns have greatly reduced the magazine's deficit. Since its initiation, *Crops and Soils* has been redesigned twice; in 1966 it was copyrighted and sale of reprints was made possible. The magazine was originally intended to better serve the farmer and his many types of advisors. In 1974 regionalization of the magazine was approved and in 1975 (October issue) regional editions were made available.

The following are the listed objectives for *Crops and Soils Magazine* as stated in 1973.

1. To be the Extension arm of the societies.
2. To present the latest research findings, with a practical approach.
3. To interpret new research information as parts of a production package or to improve environmental quality.
4. To utilize guest editorials and articles on controversial issues, as advisable.

Journal of Agronomic Education

Agronomists especially concerned with Division A-1, Resident Education, surveyed the publication situation in 1970. After considerable discussion the publication of the new journal was proposed. This was approved by ASA and in 1972 the first issue appeared. The journal was to be issued once during its first year with a final decision on the number of issues per year to be based upon the number of papers received for publication and the number of subscribers to the journal. Because of number of papers submitted and approved, the journal has been published annually to this time. Recent instructions of the editorial board relative to acceptability of manuscripts are the following:

Scope of Contributions. The Editorial Board will review: (1) reports of original research pertaining to educational concepts of resident, extension, and industrial education in plant and soil sciences; (2) analyses and syntheses of existing knowledge or research, instructional techniques and methodology, surveys of instruction, and other studies which contribute to the development or better understanding of educational efforts; and (3) reviews or digests of a comprehensive and well-defined scope; and (4) short communications and letters to the Editor. Articles may confirm and strengthen the findings of others, revise established ideas or practices, or challenge accepted theory, providing the evidence presented is significant and convincing. Manuscripts based chiefly upon personal philosophy or opinion are acceptable if they conform to the above criteria. The Editor solicits book reviews.

Journal of Environmental Quality

A "Quality of Environment" committee was established in ASA in 1969, due to the concern of members with the rapidly developing problems related to the field of agronomy generally and to soil, water, and plant research. After extensive consideration it was recommended that the *Journal of Environmental Quality* be established. This was later approved and the first issue was made available in 1972 with the journal to be on a quarterly basis. The intent of the journal is indicated in the statement from the first issue:

"The *Journal of Environmental Quality* provides an outlet for technical reports and brief reviews that are concerned with the protection and improvement of environmental quality in natural and agricultural ecosystems. Researchers in all disciplines of basic and applied science, not only agriculture, are eligible to publish manuscripts in this journal.

The purpose of JEQ is to provide a focus on environmental quality work, to provide recognition for scientists whose research is directed toward this area of investigation, and to make it possible for scientists in other disciplines to locate and recognize contributions to environmental quality more readily than in the past.

Journal of Environmental Quality readers are interested in the effect of specific practices and chemicals on environmental quality and on the quality of products in the ecosystems; pest control; management and disposal of agricultural and urban wastes; and proper land use and development."

Suggestions for Contributors to the Journal of Environmental Quality

"Contributions reporting original research or brief reviews and analyses dealing with some aspects of environmental quality in natural and agricultural ecosystems will be received from all disciplines for consideration by the editorial board. Papers may report:

1. Effects of specific practices and substances, including chemicals of agricultural or nonagricultural origin on environmental quality, ecosystems, or quality of products from natural or agricultural ecosystems.
2. Biological methods of pest control.
3. Integrated pest control programs for individual pests and ecosystems.
4. Management or disposal of agricultural wastes or use of soil for disposal of agricultural or other wastes.
5. Environmental quality aspects of land use and development.

Two main sections in the journal will be the "Technical Reports," to contain original research contributions, and the "Reviews and Analyses," to contain papers appropriate to the section title. Occasionally an editorial may be printed. Also included will be a section on "Letters to the Editor." The Editor reserves the right to extract portions of letters for publication. Critical communications will be sent to authors, and both the criticism and the author's response will be published simultaneously."

Agronomy Journal

The *Agronomy Journal* has continued to be the primary technical publication of ASA. In 1958 the circulation was 3,210 and a backlog of

200 papers. Thus the justification for the initiation of the new journal, *Crop Science*. Despite publication of *Crop Science*, there was an increase in papers for the *Agronomy Journal* in 1962 and 1963. Due to increasing problems of coordination among the three Societies' publications in 1959 the overall ASA editorial board was authorized to guide the policies of the Societies' technical journals.

Action was taken to add an abstract to journal articles in 1964 and this same year the initiation of page charges was considered. In 1965 a motion was passed to adopt the metric system for journal papers. This was to begin with manuscripts submitted after July, 1966, and published after 1 Jan. 1967. In 1971 this action was reaffirmed. In 1965 also editorial policy was amended to accept a limited number of critical review (invitational) papers in the *Agronomy Journal*. It was ruled that *Agronomy Abstracts* would not be included in Literature Cited in journals of the three Societies. While it had been considered for some time previously the three Societies agreed in 1968 that all principal plant nutrients (N, P, and K) should be expressed on an elemental basis.

During 1977 it was decided to abandon use of "light intensity" and light values based on photometric measures in Society publications. As of 1 July 1978, only radiometric unit measurements were acceptable. Consideration was given to the possibility of a journal especially for articles on tropical agriculture. The suggestion was made also for a new journal to include applied forage and range research. Favorable action on these two journals was not taken in 1977. Except for JEQ, Society journal publications are not copyrighted at present. However, consideration has been given to this matter.

In 1968 a new "Author's Guide" was prepared. The *Author's Guide and Style Manual* was first published in 1956 and was revised in 1966, 1968, 1971, and 1976. A *Monographs Style Manual* was made available in 1966 also. In 1976 a *Handbook and Style Manual for ASA, CSSA, and SSSA Publications* was completed.

In 1976 consideration was given to the question of whether material previously published in a popular publication should be accepted for the *Agronomy Journal*. Society journal editors were charged to develop uniform policy.

In recent years there has been increasing use for occasional publications to include symposia, style manuals, the *Careers in Agronomy* brochure, officer's manuals, and occasional papers of various kinds. Revision of the *Careers in Agronomy* brochure was completed in 1977. The three Societies have cooperated to publish many of them together. The ASA Special Publications series was introduced with *Food for Peace* in 1964. Increased offerings of manuscripts from foreign countries have created problems of reviewing and translating for editors.

Monographs

The policy of preparing and publishing monographs, begun with the first in 1949 has been continued. However, the first six were published by Academic Press. To date 18 have been completed... These have been either crops or soils subject areas. In 1974 it was agreed to cosponsor new monographs with CSSA and SSSA. The matter of new monographs and revision of older ones has been given continued attention. The monograph on corn was out of print for 5 years. It and those on wheat, soybeans, and turfgrass have been revised or reprinted. The following monographs were in process at the end of 1977.

Land Use Planning
Nutritional Quality of Cereals
Tall Fescue
Dryland Agriculture
Sunflower Science and Technology
Soil Nitrogen
Modeling of Plant and Soil Systems

In 1974, feasibility studies were being made for seven monographs and new proposals continue to be made. The intention of the series is to assemble and relate information in special subject matter areas of current interest. In 1974, requests were received for permission to translate two of the recent monographs into Spanish. In 1976, ten monographs were in various stages of development. A monograph *Planning the Use and Management of Land* was in preparation in 1977. This was somewhat of a parallel to the book *Soil Surveys and Land Use Planning* first published in 1966.

Symposia

Planned special topic presentations have been continued by the Society the past 20 years. Though originally begun at the annual meetings of the whole organization such programs have also been part of the meetings of the Branches. Some symposia have been held separately from the annual meetings, and papers may be published as hardcover books. Some of the national symposia presentations have been published, but not regional ones....In 1961 intersociety symposia were approved, in addition to the individual ones of the three associated Societies.

Symposia, joint sessions, and special sessions have been held by the Societies, jointly or individually as suitable. It has been difficult at times to distinguish coverage and organization of joint programs from those of special sessions or symposia.

A symposium is a series of *invited* papers which constitute a program specifically organized on a given topic. It can be sponsored by one or more divisions in the Societies. A joint program is made up of a series of volunteer papers which happen to be on the same topical area and is usually sponsored by two or more divisions of the Societies. A joint program is rarely, if ever, published in a special publication. However, symposia are often published in special publication form; the Society's special publication series was created to especially accommodate publication of symposia. Special hardcover books or proceedings of conferences might also be published....

Books

Books are usually related to symposia or special conferences held separately from the annual meetings and usually cosponsored. They are more comprehensive than other soft cover special publications.

In 1962 it was decided that the Society should not publish books if such books would be published by commercial firms.

In 1967 a proposal was made to the National Science Foundation by Division A-1, CSSA, and the Commission on Education in Agriculture and Natural Resources (CEANAR) to support writing of books in crop science. Such a series was undertaken under the title of *Foundations of Modern Crop Science* and seven volumes were planned as follows. The first of these, *Crops and Man*, appeared in 1975. Other books are in various stages of completion.

Crops and Man
Propagation of Crops
Crop Breeding
Physiological Basis of Crop Growth and Development
Ecological Basis for Crop Production
Introduction to Crop Protection
Crop Quality, Storage, and Utilization

Advances in Agronomy

Advances in Agronomy have been published since 1949. While the series would appear to be an ASA project it is under complete control of the Academic Press and is therefore a non-society publication. Officers and members of the three associated societies may make suggestions and provide help upon request. Originally suggestions were provided by the monographs committee.

Agronomy News

Originally *Agronomy News* was carried under a section heading in the *Agronomy Journal*. Beginning in 1956 it has evolved into a bimonthly publication of its own standing. Its principal purpose has been to carry current Society news items, including announcements, schedules, member changes in positions, awards, and others.

Special Publications

The Special Publication series was begun in 1962. Some of the principal types of publications so classified include papers of special Society annual programs and symposia. Career brochures, operations manuals, author's guides, special membership lists and directories, various manuals and summaries of symposia are also published by the Society but are not part of the numbered special publication series.

In 1965 a total of 15,000 copies of *World Population and Food Supply 1980* were distributed. The Society was given a citation by the "American Freedom from Hunger Foundation" for this effort. In 1968, a total of 35,000 copies of a brochure entitled *Careers in Agronomy, Crop Science, Soil Science* were printed and 15,000 copies supplied to the American Institute for Biological Sciences.

Need for publication not otherwise accommodated in the Society program has been met by this special publication classification, and a diverse use of this form has continued....

OTHER ACTIVITIES
Placement Service

The Placement Service activity began at the 1958 annual meeting. Year-around activity was initiated in 1963. In 1965, there were 210 resumes available from 41 states. In 1970, a total of 696 applicants registered at the Placement Service, and 126 positions were listed. In 1975, another 845 applicants used the service. During the 1975 annual meeting 81 agencies were interviewing and 41 departments from 33 schools were seeking graduate assistants. In 1977, a total of 709 applicants were listed by the Placement Service, 256 of whom registered at the annual meeting. Here also representatives of 47 universities, 12 government agencies, and 29 companies participated in interviews for 168 positions in crops and soils.

Manpower

Since 1959 there has been a policy committee for scientific societies in relation to manpower affairs. The ASA has been represented in this group. While conferences have not been regular, the Society has been concerned with the manpower inquiries and problems relevant to members' interests. In 1966 the Manpower Commission considered the profession of agronomy to be essential to the national welfare. Scientific Manpower Commission support has been continued by the Society.

Council for Agricultural Science and Technology (CAST)

In its meeting of June 28–30, 1971, the ASA Executive Committee took the following action:

> "It was moved that the Executive Committee recommend to the Board of Directors that ASA become an individual society member in the proposed Council for Agricultural Science and Technology..."

The Board of Directors, in its meeting of 15 Aug. 1971 approved the previous action. The Societies are founding members of CAST. In 1973, ASA had two representatives on the Board of Directors. In 1974 representatives visited in Washington to help discuss solutions of agricultural problems. Evidently CAST was actively working at that time. Discussion of implementation of CAST and the Agricultural Research Institute (ARI) relationships occurred in 1975 and also the relations of CAST and AIBS [American Institute of Biological Sciences].

The objective of CAST is to be advisory to Congress, government agencies, and the general public on national agriculture-related issues. It is the result of a meeting of the Agricultural Board of the National Academy of Sciences in late 1970. In 1972 CAST was incorporated as a non-profit educational and scientific organization.

In 1977 CAST included

   22 member societies
   83 supporting members
   20 grantors
   69 subscribers
   2,100 individual members

By this year CAST had prepared 69 reports, 4 special publications, and 6 papers.

American Registry for Certified Professionals in Agronomy, Crops, and Soils (ARCPACS)

The suggestion for professional certification of agronomists was considered in 1968 by the Organization and Policy Committee and tabled until further study could be made. In 1969 a special committee was appointed to develop specifications and standards for certification, and in 1972 a proposed procedure was presented. Establishment of a certification program was recommended in 1974 by a study committee.

The operation of the American Registry of Certified Professionals in Agronomy, Crops, and Soils (ARCPACS) was begun in 1977. The stated purpose was, "to develop standards and procedures for certification of persons qualified as professionals in agronomy, crops or soils, and to maintain and publish a registry of persons so qualified." A Board for Certification passes on each applicant. Three sub-boards, one for each discipline, comprise the total ARCPACS board. ARCPACS is a non-profit corporation operated in cooperation with the three Societies. It is supported equally by the three Societies on a loan basis. The Director is an ASA staff member and the program is housed in the Headquarters office.

In 1960, SSSA members voted for certification of soil scientists, consideration beginning in 1953. A soil scientist certification program for undergraduates was begun in 1964.

The areas of specialization to be certified by ARCPACS are Agronomist, Crop Scientists, Crop Specialist, Soil Scientist, Soil Specialist, and Professional in Training for those applicants who do not have the needed years of experience to qualify for full certification. A registrant may be certified in more than one classification. Registration is renewed annually and is subject to review.

Archives

The Western Society of Soil Science raised the question of the establishment of archives in 1972. A committee to consider the desirability of establishing an archives program was appointed in 1973. It was recognized that the problem of maintaining records or archives was of concern for the three Societies and for the regional branches. This matter was under study until 1977 and the Iowa State University library and that of the University of Wisconsin offered to store archive records for the three associated Societies. The latter option was chosen in 1977. The guidelines for establishing the archive holdings are still to be defined.

The Headquarters office, upon agreement and consultation, will provide storage for essential records of regional branches.

Agronomic Science Foundation

In 1964, the Organization and Policy Committee was asked to consider the creation of a foundation to accept gifts. During 1965, Articles of Incorporation were formulated and a draft authorized by the committee. The final Articles of Incorporation were approved in 1966 and the Agronomic Science Foundation was incorporated in 1967. A request was made to the Internal Revenue Service for gifts to be tax-deductible. The first Board of Trustees was appointed and in 1968 the Foundation was ready to move into an operational program. It has served as a receiver of gift funds and as a project and support planning group since that time. Five major uses for ASA funds were identified early by the Agronomic Science Foundation as:

1. Development of the Agronomic Science Foundation activity in developing countries.
2. Establishment of career promotion programs in agronomic sciences.
3. Establishment of a scholarship program in the agronomic sciences.
4. Financial assistance to educational programs in the agronomic sciences.
5. Financial support of specific projects submitted by ASA, CSSA, and SSSA or their members.

The executive vice president has served as director since the beginning.

The Articles of Incorporation were published in the *Agronomy Journal* 60:133–134 (1968).

Visiting Scientist

A proposal was developed to have a Visiting Scientist program in 1962. This program was to make available upon request, agronomic scientists who were primarily research workers, to present technical lectures on college and university campuses. The primary purpose was to acquaint students with better insight of agronomic and related research and to stimulate interest.

A special Visiting Scientist panel was established and plans and rules for the program developed with the executive vice president as director. Campus groups had several choices as to subject matter presented. The program was initiated in 1963 with the financial support of the National Science Foundation. This continued until 1971–72 when it became necessary to discontinue the program because of discontinuation of support.

Registration of Field Crop Varieties

Registration was begun very early for some crops, primarily as an effort of USDA. In 1958, a cooperative understanding was developed between the Agricultural Research Service (Crops Division) and ASA. In 1961, a committee was appointed to work with the seed industry concerning problems of variety registration. The responsibility of ASA for variety registration was terminated and transferred to CSSA in 1965. Since that time CSSA has carried on that activity.

Other

In 1962, publication of M.S. and Ph.D. theses titles and authors was begun.

The Society provided letters to the President's Council on Environmental Quality and to the Director of the Environmental Protection Agency stating the Societies' concern and interest in environmental problems and offering help of the membership, in 1974.

A resolution was drafted citing the accomplishments and contributions of agriculturists to the status of the United States in our bicentennial year. Also the bicentennial year was used to publicize the advancements in agriculture.

In 1976 four areas were identified as of major concern of agronomic scientists, including:

1. Potential for energy from renewable resources.
2. Food and Drug Administration regulations on toxicity levels in new plant varieties.
3. Land use planning.
4. Pesticides.

It was proposed in 1975 that a survey be made concerning foreign language requirements for graduate students. Application and use of foreign languages have become of major concern to many agronomists.

In 1977 it was suggested that the Society offer help to USDA in the selection of members for grant proposal review panels.

In 1977, it was proposed that a "Public Service Associate" or "Congressional Fellow" program be undertaken. This would allow young workers, eligible for sabbatical leaves and having interest, to spend a year in Washington improving acquaintance with governmental operation and associating with Congressional Aides. Society members have been concerned that closer involvement, and participation in activities affecting agronomists especially, be developed.

Other aspects of the Society program have been considered elsewhere.

International Programs

Interest in international aspects of the ASA program began to be more active in 1962 when a sectional program on "Pan-American Agronomy" was agreed upon. In 1965 travel and a professional visit by the executive secretary to Latin American countries was approved. A principal visit was made to the Association Latinoamericana de Fitotecnia (ALAF), with Headquarters in Cali, Colombia. ALAF was founded in 1961. This visit was supported by ALAF and the Rockefeller Foundation. In 1969 there was a suggestion of a joint meeting of ASA arid ALAF at Bogota, Colombia. A special committee was appointed to develop cooperation with ALAF.

A committee on "Foreign Relations and Activities" was established in 1965, and in 1966 there was consideration of efforts to give more emphasis to international agriculture. An International Agronomy committee was appointed in 1967. The suggestion was made also that professional societies be encouraged in other countries. In 1968 a total of 400 copies of a resolution prepared by the committee was sent to key people in government, universities, foundations, academies, and industries concerning food supplies and needs. In 1971 there was action to facilitate provision of printed material from the Societies to foreign countries. The Norman E. Borlaug Book Fund was established in that year. Also, reprinting of foreign editions of ASA monographs or other books, in foreign languages, was approved.

The possibility of translating publications of East Asian countries into English was considered in 1972. Funding was recognized as a problem. In 1973 collection and re-distribution of donated books to various countries was discussed.

In 1975 a committee was appointed to investigate liaison with comparable societies in other countries. The associated societies discussed a program to provide Society journals and books to libraries in developing countries.

In cooperation with CIAT the arrangement was made for a representative to visit the ASA Headquarters office in 1977 to study the distribution of scientific information in a popular format.

Meetings

National meetings of ASA have been continued to be held annually since 1958....[2] Beginning in 1961, the national meetings have been well distributed over the country and continual efforts have been made

---

[2] *A list of meeting locations and themes is included on the accompanying CD.*

to continue this pattern as accommodations have been available. The growth of the Societies and the large group attending meetings has complicated arrangements. Fortunately many university campuses have greatly expanded facilities and are now capable of scheduling the meetings in summer months. Locations of national and regional meetings are coordinated. Branch meetings have also been held regularly...

In 1977, a total of 3,500 members and guests were in attendance and 1,130 papers were presented at the Los Angeles meeting. Horizontal and vertical growth of the Societies has been accompanied by numerous special programs, including symposia, invited lecturers, and conferences.

Attempts have been made to have meetings jointly with other societies, especially the American Society of Horticultural Science. Such meetings have been considered desirable but divergence of interests and difficulties in scheduling have prevented such meetings thus far.

Beginning in 1962, a central theme was chosen for the annual meeting of the societies by the ASA president-elect and with the approval of the executive committee. Special publications relevant to the previous themes have been made available by ASA....

They indicate the concentration and emphasis of the programs and the trends over the past 20 years. More discussion of the programs is included elsewhere.

COOPERATION

It is of interest and significance to note the relations of ASA to other organizations directly and broadly related to agriculture and to biology. Society individual representatives and committees are related to the following:

> American Forage and Grassland Council
> Assembly of Life Sciences, NRC-NAS
> Agricultural Research Institute, NRC-NAS
> Collaboration with American Society of Agricultural Engineers
> Soil Conservation Society of America
> Society for Range Management
> Weed Science Society of America
> Association Latinamericana de Ciencias Agricolas (ALCA)
> National Science Foundation
> American Society of Animal Science
> American Dairy Science Association
> Generally Regarded as Safe (GRAS)
> American Institute of Biological Sciences (AIBS)
> American Meteorological Society
> Entomological Society of America
> Genetics Society of America
> Society of American Foresters
> American Society of Horticultural Science
> American Phytopathological Society
> American Agricultural Economics Association
> Council for Agricultural Science and Technology (CAST)
> International Turfgrass Society

The CSSA and SSSA may have additional associations with other societies. A brief chronological review of some of the cooperative relations follows.

In 1959 it was recommended that ASA establish full membership in AIBS. Reasons and alternatives were discussed. Dues required created a problem. This was considered further in 1961 and 1962. The ASA became an adherent Society of AIBS in 1965 following further consideration in 1963. In 1972, membership in AIBS was questioned as it was considered to be too political, however membership has been continued to the present.

Numerous cooperative relationships have existed with other societies. In 1959, ASA participated in a conference with the American Meteorological Society and eight other societies.

In 1961 there was an effort to determine draft policy by the Society Committee on Agronomic Manpower Resources. In 1965 this Committee was discontinued. In 1968 a policy committee for scientific agricultural societies including ASA, met with the Manpower Commission concerning the proposed elimination of graduate student deferments.

A joint meeting was held with the American Society of Range Management in 1964.

In 1965 cosponsoring of conferences, seminars and symposia, and publishing of resulting papers was approved. Guidelines were developed in 1968. ASA cosponsored a Resident Instruction Conference and Program in 1977 with the Commission on Education in Agriculture and Natural Resources, which operates within the Division of Biology and Agriculture of the National Research Council.

Consideration was given to activities of the League for International Food Education (LIFE), founded in 1968, in 1971. In 1972, ASA became a member of this organization.

In 1969, ASA participated in the International Science Fair. The Society cosponsored the Third International Seminar of Hydrology Professors in 1971. Action was taken to participate in the World Environment and Resource Council in 1971.

The Society became interested in Generally Recognized as Safe (GRAS) group in 1972. This interest stemmed from the FDA concern about the possible toxicity characteristics of forthcoming new varieties; GRAS was an investigational group in FDA.

In 1972 the Peace Corps asked the Society for help in recruiting agronomists. Action was taken in 1973 to cooperate with the Peace Corps in procurement and screening of personnel for its use.

In 1972, ASA rejoined the American Forage and Grassland Council and increased cooperation with ASA was sought by the Weed Science Society of America. In 1972 the Society agreed to cosponsor and publish the proceedings of the International Turfgrass Society.

It was voted to become a member of The Institute of Ecology (TIE) in 1973.

The ASA has been cooperating in the program of Section O of AAAS as a voting member for many years. In 1973 it was considered that participation in Section G, Biological Sciences, be included. There has been association with four other divisions.

The ASA participates in the activity of ARI, which in 1973 included 143 members. These were

> 88 industries and industrial organizations
> 38 state institutions and organizations
> 13 U.S. government agencies
> 4 scientific societies (including ASA)

It was agreed by ASA in 1975 to cosponsor a "Fourth Conference of the International Association on Mechanization of Field Experiments," with the American Society of Agricultural Engineers. It was agreed to support that Society also in developing a "Conservation Tillage Handbook." A National Soil Erosion Conference was cosponsored in 1976.

The ASA has a working relationship with the Council on Soil Testing and Plant Analysis though not affiliated with it.

There has been continued cooperation with such organizations as the Association of Official Seed Certifying Agencies (AOSCA) (formerly International Crop Improvement Association), the association of Official Seed Analysts (AOSA), the Crop Quality Council, the American Seed Trade Association and many others, over a long period.

Agronomy-Industry luncheon meetings have been held during the annual meetings since 1948, and industry members of the Society have participated in all activities of interest.

Possibilities of joint meetings with animal science societies have been considered. Closer liaison with the American Agricultural Economics Association was suggested in 1977.

It is difficult to list all of the cooperative relations which have occurred with the many organizations and societies over the past 20 years. Those included indicate the growth of the Society program in activity and diversity.

Building the ASA Headquarters in Madison

Dale Smith
Professor emeritus, Department of Agronomy, University of Wisconsin-Madison.

Several people in the University of Wisconsin (UW) Agronomy and Soil Science Departments were heavily involved in building and expanding the Headquarters for the American Society of Agronomy (ASA) in Madison, WI. This was particularly the case for the Building Committee appointed in January, 1961, by ASA President Rodney Bertramson. This committee was composed of Ed Engelbert (Chairman), Marvin Beatty, and Arthur Peterson of the UW Soil Science Department, and N.P. Neal, Donald Peterson, and Dale Smith of the UW Agronomy Department. The duties of this committee included the inspection of numerous possible building sites in Madison, the selection of an architect and contractor, approval of building plans and materials, and choice of draperies and furniture. This committee was active from the spring of 1961 until it was unofficially disbanded in 1969.

The trail of an ASA Headquarters in Madison was a long one. An ASA Committee on New Publications, chaired by K.S. Quisenberry, recommended in 1947 that a permanent Headquarters for the ASA be established in a small to moderate size city, and preferably in the same city as the publisher. A city in the central part of the country was preferred. Several months later, it was announced that Lawrence G. Monthey had been appointed Editor-Secretary of ASA (Monthey had received his B.Sc. in Soils at Wisconsin in 1940 and his M.Sc. in Agronomy at Iowa State in 1947). His first duties were to establish a Headquarters and start a popular magazine (*What's New in Crops and Soils*). Pending the establishment of a permanent Headquarters and the launching of a new publication, Dr. G.G. Poehlman would continue to handle the business and financial affairs of the Society in Morgantown, WV, and Mr. J.D. Luckett the technical publication at Geneva, NY. Mr. Monthey established the Headquarters in Madison, WI in January, 1948, in a rented building at first at 1910 Monroe Street, but soon moved to 2702 Monroe Street where it remained on a tentative basis until being made permanent in 1960.

In 1960, a Central Office Site Location Committee, under the chairmanship of W.H. Pierre, was appointed by ASA. After considerable study, it submitted the names of three cities to the membership for a vote by mail ballot. The ballot on whether to build a permanent Headquarters in Columbus, OH; St. Paul, MN; or Madison, WI was sent to all voting members of ASA in the fall of 1960. The final vote was: Madison, WI—1435; Columbus, OH—254; and St. Paul, MN—171. In the meantime, Lawrence Monthey resigned as Executive Secretary effective October 1, 1960, to become an environmental research specialist with UW Extension. Matthias Stelly became Executive Secretary effective April 1, 1961. (Stelly graduated from Southwestern Louisiana Institute and then received an M.Sc. from Louisiana State and a Ph.D. from Iowa State in 1942, both in soil fertility. He came to the ASA position from the staff of the Soils Department of Louisiana State University.)

After the Building Committee of Engelbert (Chair), Beatty, Arthur Peterson, Neal, Donald Peterson, and Dale Smith had been appointed by ASA President Rodney Bertramson in January, 1961, a site in Madison, WI became the first order of business. After looking at several sites around Madison, a 1.59-acre piece of land was purchased on March

Ground-breaking for the ASA Headquarters. (Front, left to right): B.R. Bertramson, Emil Truog, L.F. Graber, G.F. Sprague, and Architect H.R. Ames. (Back, left to right): W.L. Nelson, SSSA President; J.R. Cowan, CSSA President; C.A. Black, SSSA Vice President; G.W. Burton, ASA Vice President; M. Stelly, Executive Secretary; L.E. Engelbert, D.R. Peterson, A.E. Peterson, Building Committee; J.H. Crissinger, Contractor.

Unpublished report on file
Date of submission estimated to be the early 1970s.

31, 1961 for $43,719. It was a triangular piece of land in west Madison between Segoe and Odana Roads. The land was gently sloping from Segoe Road on the north to Odana Road on the south. On Segoe Road, the land was across the street from the Westgate Shopping Center and an elementary school. On Odana Road, it was across from a residential area and not far from a park and golf course. The Building Committee was fortunate to get this site since an apartment complex on the east edge of the parcel of land was considering the area for more apartment buildings.

After purchase of the land, the Building Committee went to work with the architectural firm of Ames, Torkelson, and Nugent to design a Headquarters building. The July/August 1961 *Agronomy News* shows the plans for a one-story, fire-proof, steel and masonry building that would face Segoe Road. The natural slope permitted the development of an exposed basement open toward Odana Road on the south. The total area of the two floors, ground and main, was about 5840 sq. ft.

The lowest bid for construction came from the J.H. Crissinger Construction Co. The winning bid was for $115,000 for construction and site improvement, with $87,083 ($14.9l/sq. ft.) for building construction only. The contract contained a clause allowing a maximum of a $3,000 cost overrun.

A ground-breaking ceremony was held on November 3, 1961, and construction started within the month. ASA President Rodney Bertramson and three ASA past presidents—L.F. Graber, G.F. Sprague, and Emil Truog—drilled holes with soil augers into the Miami silt loam soil to break the ground for construction. They performed the symbolic function for all the members of ASA and they represented much of the history of the three Societies. Emil Truog of the UW Soil Science Department was ASA President in 1926 and of SSSA in 1954. L.F. Graber of the UW Agronomy Department was ASA President in 1950. G.F. Sprague of the Illinois Agronomy Department was ASA President in 1960. Both were former Chairmen of the ASA Crops Section that later became the CSSA. The ceremony was also attended by members of the Building Committee, the combined Budget and Finance Committee, and other dignitaries (Jan./Feb. 1961 *Agronomy News*).

While constructing the Headquarters building during late 1961 and early 1962, the Building Committee worked day to day with Matthias Stelly, the Executive Secretary of ASA. Some of the problems were small indeed. At times, Stelly was testy when our decisions did not go his way. However, Ed Engelbert, Chairman of the Building Committee, did a masterful job of moderating the controversies that arose with Stelly and/or the contractor. In the end, a very excellent Headquarters building was constructed on a quite spacious piece of land.

The building was completed in July, 1962, and the Headquarters staffs of the ASA Societies moved into the new Headquarters at 677 South Segoe Road on Madison's west side on July 21, 1962. A grass lawn had been established in late May by sodding the erodable slopes and seeding the remainder. This work had been contracted to Dr. J.M. Scholl, Professor of Agronomy at the UW. The shrubbery was then planted in the spring of 1963 based on a landscape plan developed by the Building Committee.

The formal dedication of the Headquarters building took place on October 29, 1962. Dr. D.C. Smith of the UW Agronomy Department was chairman of the Dedication Program Committee and presided at the ceremony. The featured speaker was ASA President M.B. Russell of the University of Illinois. Madison Mayor R.F. Reynolds gave remarks on behalf of the City of Madison. Executive-Secretary Matthias Stelly described the functions of the Headquarters office and introduced individuals who contributed to the building project. More than 100 people attended the program and inspected the new building. It was a 19-month period from the purchase of the land to the dedication of the building (Agron. J. Nov./Dec., 1962, pages 522-A, -B, and -C).

The Societies office during construction, upon completion, and today.

In 1966, the Society approved the construction of a 3,276 sq. ft. addition to the Headquarters building in Madison (one-half the area of the original building). The building committee for the original building also was involved in the addition, except for Arthur Peterson, who was on a foreign assignment with the Ford Foundation. The first addition was begun in the spring of 1967 and was moved into by the Headquarters staff in December, 1967. Ames, Torkelson, and Nugent again were the architects. The contractor was Joe Daniels Construction Co.

The ASA Building Committee was unofficially disbanded in 1969 when the Chairman, L.E. Engelbert, left in June 1969 for Brazil to head the University of Wisconsin-Madison/AID Program at the Federal University of Rio Grand do Sul in Porto Alegre, Brazil. I kidded Dr. Engelbert that he was taking the Brazil position to get away from the problems involved with the Building Committee.

Acknowledgements

The writer wishes to thank Mr. David M. Kral, Associate Executive Vice President of ASA.

Editor's Note

During the extremely rapid growth period and increase of Headquarters activity and staff, space again became limiting in 1974. A 6200 sq. ft. addition to the Headquarters building was authorized by the board in 1975, and the new addition was completed and occupied on 13 October, 1976. In 1984, the building was remodeled again, and a new entrance and reception area were added.

Development of the American Society of Agronomy 1978–2007

The American Society of Agronomy (ASA) was formed at a meeting on December 31, 1907 held in the Department of Botany at the University of Chicago. The constitution and bylaws were adopted with the objective to "Increase the dissemination of knowledge concerning soils and crops and the conditions affecting them." Presentation of papers was given a high priority at the first meeting. With this objective in mind, ASA has been at the forefront of scientific and educational organizations disseminating science via publishing peer-reviewed journals, books, and other educational aids; conducting international meetings, promoting professionalism, developing relationships with universities, government, and other organizations to enhance science; and continuously seeking to add new programs and activities.

The purpose of this chapter is to provide a summary of the changes and the development of the ASA from 1978 to 2007. During this 30-year period the Society has grown extensively and the scope of activities has been greatly broadened. A major development recently has been the reorganization of ASA and its relationship with the Crop Science Society of America (CSSA) and Soil Science Society of America (SSSA), the Agronomic Science Foundation (ASF), and related activities. The history and development of the ASA during the first 75 years has been described by previous authors (Lyon, 1933; Laude, 1962; Smith, 1980), and their accounts were presented in the preceding chapters.

The operations and functions of the Society are not dependent only on the contributions of elected officers and staff. In addition to their efforts, hundreds of members and outstanding workers contribute. Since it is not possible to give credit to all originators of ideas, names are being omitted in most instances. The activities, procedures, and programs of ASA, CSSA, and SSSA are markedly interrelated. Also, in reviewing the history of ASA, the development of the other two Societies is highly relevant.

ORGANIZATION

Membership in ASA reached its peak of 12,781 in 1986. After a relatively stable period until 1992, membership numbers have declined for various reasons. However, the Society is involved in a marked increase in activities and programs. To increase identity for all the associated organizations, a major activity has been the reorganization and relationship of the three Societies. The ASF continues to be a separate organization serving all three Societies. New bylaws and board structure approved in 2005, resulted in reducing the board of directors from 45 to 16. The primary purpose of the reorganization was to establish a board and executive committee of a more manageable size, thus achieving improved continuity, communication, and responsiveness. For additional information refer to the ASA webpage at www.agronomy.org. At this site one can find detailed information about various aspects of ASA.

Dwayne A. Rohweder
Emeritus Professor of Agronomy
University of Wisconsin, Madison

Robert F Barnes
ASA–CSSA–SSSA Executive Vice President, Retired

David M. Kral
ASA–CSSA–SSSA Associate Vice President, Retired

Calvin O. Qualset
Research Professor, Professor and Director Emeritus
University of California, Davis
Chair, ASF Board

From left, Dwayne Rohweder, Bob Barnes, and Dave Kral meeting at the ASA office to check facts for this historical summary. 2007.

Corresponding author
D. Rohweder, Emeritus Professor of Agronomy Extension, University of Wisconsin-Madison (email: darohwed@wisc.edu).

The authors are indebted to members of the Headquarters staff for assistance in providing information for this history.

Board of Directors and Executive Committee

The 16 members of the restructured ASA Board of Directors are: ASA President, ASA President-elect, ASA Past President, nine division representatives, one member elected from the International Certified Crop Adviser program, and the following ex-officio and nonvoting members, the ASA Editor-in-Chief, the Chair of the Agronomic Science Foundation, and the Executive Vice-President. The Executive Committee is composed of the ASA President, ASA President-elect, and ASA Past President. As of 2005, the CSSA and SSSA officers are no longer part of the ASA Executive Committee or Board. With the smaller Executive Committee and Board, more involvement of board members in key ASA activities such as budget, outreach, policy, rapid response to policy issues, program development, and well as member recruitment and retention are envisioned. In addition, the organizational structure will focus specifically on ASA programs and services unique to agronomy.

Management Group

Since CSSA and SSSA representatives are no longer on the restructured ASA board, a nonprofit corporation known as the Alliance of Crop, Soil, and Environmental Science Societies (ACSESS) will be created to be the management and business unit for ASA, CSSA, SSSA, and ASF. In addition to providing services, the organization also should provide a forum to discuss and implement issues that integrate crops, soils, and environmental sciences and areas that cut across ASA, CSSA, and SSSA. The organization will focus on ASA, CSSA, and SSSA and is expected to allow for additional societies in the future. Bylaws and articles of incorporation are currently being drafted, with a target date of January 1, 2008 for ACSESS to begin management services. The Board also approved the title change of the Executive Vice President to the Chief Executive Officer, effectively immediately in 2007. For many years an Associate Executive Vice President handled the day-to-day activities as well as the majority of personnel management. A Chief Operations Officer position has been created and is responsible for the day-to-day activities, including personnel management for the majority of staff.

The primary mission of disseminating agronomic science via publishing peer-reviewed journals and conducting international meetings will continue. Expanded offerings will be made to enhance value to members and participants, and the Society will continue to explore the addition of new programs and activities.

Committee Structure

In 1977, there were 73 ASA, 41 ASA–CSSA–SSSA (ACS), 4 AC, and 2 SC committees operating. In 1996, this number had increased to 94 ASA, 48 ACS, 11 AC, and 6 AS committees. Both CSSA and SSSA have their own committee structures as suited to their individual Societies and their association with ASA. Currently, the work related to the activities and functions of ASA have been divided into 10 major areas:

Activity	*Number of Committees in 2007*
Organization	18
Nominations	10
Awards	25
Operations and finance	4
Publications	7
Professional advancement	7
Collaboration	6
Meetings	3
Scientific Affairs	1
International Activities	0

Today's ASA leadership (From top left): 2007 President Jerry Hatfield, President-Elect Kenneth Moore, Past President David A. Sleper, and 2008 President-Elect Mark Alley.

Divisions of ASA

In 2007, ASA has nine broad subject matter Divisions through which programming is performed:

A-1 Resident Education
A-2 Military Land Use & Management
A-3 Agroclimatology & Agronomic Modeling
A-4 Extension Education
A-5 Environmental Quality
A-6 International Agronomy
A-7 Agricultural Research Station Management
A-8 Integrated Agricultural Systems
A-9 Professional Practitioners

The subdivision (A-1a) of Division A-1 for student activities was discontinued in 2004, and a new student organization, Students of Agronomy, Soils, and Environmental Sciences (SASES) was developed. The SASES is affiliated with all three of the Societies. Since 1977, Division A-2 changed its name from Land Use and Management to Military Land Use and Management (1984) and Division A-3 changed from Climatology to Agroclimatology and Agronomic Modeling (1980). Division A-7 was permanently added in 1984. Division A-8 was first called Plant Science Applications, and in 1995, it became permanent and the name was changed to Integrated Agricultural Systems. Division A-9 was established as a steering committee, Professional Practitioners, in 2000. In 2002 Professional Practitioners became the permanent A-9 Division. The A Divisions are a mixture of integrative subject matter areas (A-3, A-5, A-8, and A-9), management (A-2 and A-7), and crosscutting areas (A-1, A-4, and A-6) that serve all three Societies.

Regional Branches

The overall ASA–CSSA–SSSA membership by region in 2006 was: North Central 31.5%, Northeast 9.5%, Southern 24.7%, Western 17.9%, and international, including Canada, 16.2%. The four regional branches of ASA have long been in existence, with interest and support varying region by region. Annual meetings of branches have usually been held. Regional branches continue to have a loose liaison with ASA. There are four major ASA branches: Northeastern, Southern, North Central, and Western. The Northeastern branch recently has served as a branch of ASA and SSSA, and in November 2005 it was officially approved as a CSSA branch as well. The Western Regional Branch of the Society is composed of the Western Society of Crop Science (WCSC) and the Western Society of Soil Science (WSSS). However, both organizations are independent of each other and usually meet separately. The WSSS usually meets in conjunction with the Pacific Division of the American Association for the Advancement of Science. The two organizations meet jointly periodically. Each branch has its own programming and emphasis. The North Central Branch changed its emphasis to practicing agronomists in 1994.

Support for the branches has been available from ASA. Society assistance has been given in the publication of programs and abstracts of papers of regional meetings. Officers of the regional branches have met to confer on the relationship of branches to the national society, resulting in some coordination. However, as of 2005, there are no regional representatives on the ASA Board. Most of the activity of ASA is on the national level.

Early in the history, state chapters were encouraged to stimulate ASA membership and to develop interest in the agronomic profession. However, over the years they have gradually disappeared and today little individual state activity exists.

Budget

Budgetary statements have been prepared by the Budget and Finance Committee and presented to the Board. Budget committees have spent considerable time in association with Society officers and the Headquarters office staff in consideration of all aspects of the budget, including the disbursement of income and conformity to Internal Revenue Service requirements for tax-exempt status (501.C.3 scientific, educational). Investment of surplus funds, utilization in equipment purchases, and establishment of pension reserves also have been included in the consideration of various committees. Sharing of costs in operations and projects of mutual interest has been established.

MEMBERSHIP

During the decade 1977 to 1986 membership increased 32% (from 9653 to 12,781). This 12,781 membership number was the all-time high. The 10,000th member, Donald J. Dorsett, was enrolled in October 1978. In the next decade (1986–1996), ASA membership held relatively steady, dropping only 9% (12,781–11,807). However, during the third decade (1996–2006), ASA membership dropped 29% (11,807–8,378). Multiple factors caused this decline in membership, including:

- The organization of new, competing scientific organizations.
- Diversification of traditional university departments into more basic areas.
- The dramatic number of retirements (post WW II era hires).
- Scientists in the crop and soil sciences may choose among numerous, more specialized societies that may better represent their areas of interest.
- Many scientists today do not join as many societies as did the retiring scientists.
- A decline in university student populations entering the fields.
- A change in scientific research funding.
- Today's ASA membership is not automatic when one joins CSSA or SSSA, as it was in 1977.

However, international membership has held relatively steady during the past 25 years, at 15 to 20% of the total membership. In 2005, a 3-year trial was launched with a Developing Countries Membership. It is open to scientists and students in the lowest two economic tiers of the World Bank country list, with the stipulation that both residency and citizenship in an eligible country are required. It includes membership in one Society and online access to one journal and *CSA News*. In addition, via ASF, these scientists also may apply for financial assistance of one-half the cost of the Developing Countries' membership. In 2006, approximately 110 members participated via the Developing Countries Membership.

The number of membership options in ASA has been reduced from 12 in 1977 to eight in 2007. Current membership options are: Active, Affiliate, Emeritus, Sustaining, Graduate, Undergraduate Student Affiliate, Subscriber, and Certification Member. Each sustaining membership has at least one identified member representative with full membership privileges. In addition, there is Honorary Membership, which is considered an award, and International, which is a subcategory of other membership options. Life and Associate membership options have been eliminated.

Awards

The growth, increased activity, increased financial capability, and outside support of ASA has resulted in a substantial expansion of the award program during this period. Awards designated during the 1978–2007 period are: Werner L. Nelson Award for Diagnosis of Yield Limiting Factors (1990), Harry J. Larson/Yara Memorial Scholarship (1998), Hank Beachell Future Leadership Scholarship (1998), Monsanto Professional Certification Service Award (2003), Carl Sprengel Agronomic Research Award (2003), Environmental Quality Research Award (1984), J. Fielding Reed Scholarship (1997), Frank D. Keim Graduate Fellowship (2001), Agronomic Industry Award (1988), Deere and Co. Graduate Fellowship (2001), E.T. & Vam York Distinguished ASA Lectureship (2003), and Professional Practitioner Award (2004). Continuing ASA awards are: ASA Fellow, Agronomic Extension Education Award, Agronomic Resident Education Award, Agronomic Service Award, International Service in Agronomy Award, Syngenta Crop Protection Recognition Award, and ASA Honorary Membership. Student awards are discussed in the chapter on resident education (Milford and Smith, 2007, this publication). See the accompanying CD for a list of award recipients.

Fellows

The first Fellows in the American Society of Agronomy were named in 1925. Including the year 2006, a total of 1569 Fellows have been named. Beginning in 1976, SSSA initiated its own Fellows program. CSSA initiated its own program in 1985. In 2005, the number of years of membership required for eligibility to be nominated as an ASA Fellow was reduced from 10 to 7 years of active membership.

Current criteria and weighting for nomination as Fellow are as follows:

Criteria	Points
Personal Achievements	15
Professional Achievements	60
Professional Contributions	15
Letters of Support	10
Total Points	100

Today's Societies' Office staff is divided into nine departments: (from top left): administrative, meetings, publications, science policy, ASF, IT and operations, finance, certifications, and membership. Dr. Ellen G. Bergfeld (top left) is the Chief Executive Officer. Staff (from top left): Ellen Bergfeld, Cathy Goudreau; Linda Nelson, Pat Scullion, Stacey Phelps, Keith Schlesinger; Carrie Czerwonka, Fran Katz, Liz Gephardt, Rebecca Funck, Kevin Ducey, Matt Nilsson, Sue Ernst, Lisa Al-Amoodi, Meg Ipsen, (Ann Edahl, publications, not shown); Karl Glasener, (Caron Gala, science policy, not shown); Paul Kamps; Keith Lovejoy, Elizabeth Tucker, Tom Moeller, Johanna Cherry, Ian Popkewitz; Priscilla Westra, Chris Meinholz, Lynne Navis, Michela Cobb, Penny Magana, Kathy Heiderscheit; Laurie Karr, Mary Jo Rowbottom, Michele Lovejoy, Luther Smith, Darla Steinborn, Lorene Peterson; Leann Malison, Melissa Fall, Sara Uttech, Valette Piper-Bledsoe, Susan Chapman, and Nellie Mitchell.

HEADQUARTERS STAFF

Since 1961, increasing activities, work, and responsibilities have required additions to the Headquarters staff. Lawrence (Larry) G. Monthey was the first staff Executive Secretary of the Society and served from 1948 until 1960. He established the Headquarters office in Madison in rented space.

Headquarters office administrators that served during the last 30 years include the following:

Mathias (Matt) Stelly	
Executive Vice President	1 Apr. 1961–31 Dec. 1981
Rodney A. Briggs	
Executive Vice President	1 Oct. 1982–30 Nov. 1984
Robert F Barnes	
Executive Vice President	30 Sept. 1986–17 Feb. 1999
John Nicholaides III	
Executive Vice President	1 July 1999–2 Jan. 2002
Ellen G.M. Bergfeld	
Executive Vice President/ Chief Executive Officer	1 July 2003–present
David M. Kral	
Associate Executive Vice President	1 June 1967–1 Jan. 2005
Acting Executive Vice President	Jan.–Sept. 1982, Dec. 1984–Sept. 1986, Jan. 2002–June 2003
Frances Katz	
Chief Operating Officer	April 2007–present

ASA PUBLICATIONS

The ASA currently publishes three research journals, monographs, a news magazine, and a professional magazine. The journals are *Agronomy Journal* (AJ) published by ASA; *Journal of Environmental Quality* (JEQ) published by ASA, CSSA, SSSA; and *Journal of Natural Resources and Life Sciences Education* (JNRLSE) published by ASA along with nine cooperators. Starting in 2002, selected papers from each issue of the journals are highlighted, with news releases prepared and distributed through a service of the American Association for the Advancement of Science to more than 4000 international science media. Articles also are promoted on the Society web pages, in *CSA News*, and the News Flash.

Agronomy Journal (AJ)

The AJ has continued as the primary technical publication of the ASA (Pearson et al., 2007, this publication). The Centennial Issue of *Agronomy Journal*, Volume 100, will be published in 2008. In 1978, the circulation of AJ reached a total of 8571 (6485 members). Circulation reached a high of 9607 (7742 members) in 1985. Circulation then started to decline, with only 2783 (2175 members) subscribing in 2006. The membership requirement for purchase of AJ was discontinued in 1997. Total pages increased from 1096 in 1999 to 1828 in 2004 and dropped slightly to 1692 in 2005. The number of articles published increased from 142 to 208, and then reduced back to 188 in 2005. Average pages per article increased from 6.6 to 8.1 with 26% international papers.

Style continued to evolve during this time. The journals adopted the SI system (Système International d'Unités) of reporting units in 1982–1983. Revised style manuals were published in 1984, 1988, 1989, and 2004. The 2004 revision was published online only and is updated as needed. The AJ Editorial Board divided the table of contents into categories. The categories "Crops" and "Soils" first appeared in the January-February 1990 issues. Categories continually change with the nature of the research published. Today numerous flexible categories are used to better help readers identify the section(s) of interest in the journal. Page charges for published articles have changed every few years to reflect publication costs. Beginning in 2004, the traditional page charges were replaced with a publication fee. The publication fee on a per-manuscript basis was $450 for members and $700 for nonmembers. In 1999, AJ was published with a new cover featuring color pictures and began accepting production agriculture papers since JPA was slated for termination.

The AJ has been published electronically by HighWire Press, Palo Alto, CA since 2001. The online version is available at http://agron.scijournals.org/. As of September 2005, the average number of hits to the home page of AJ was 9196. Subscribers can choose print, electronic, CD, or any combination of the categories. The distribution of the 3085 AJ subscriptions in 2005 was: print 1797, print–electronic 249, print–CD 271, print–CD–electronic 118, CD 152, electronic 246, and electronic–CD 270. Print subscribers still totaled 2417, or 78%. Beginning in mid-2005, AJ was published continuously; manuscripts are posted online each month with fully citable reference information. Manuscript Tracker was implemented for submitting, registering, reviewing, revising, and tracking manuscripts in 2002. A seven-CD set of the back issues (Vol. 1–93, 1908–2001 of *Agronomy Journal* became available for purchase in 2004. The CD set for AJ also included the *Journal of Production Agriculture* and the *Journal of Natural Resources and Life Sciences Education*. In 2008, all of the back issues of AJ are being made available online by HighWire Press at the journal site, http://agron.scijournals.org/.

Journal of Environmental Quality (JEQ)

The JEQ, published by ASA, CSSA, and SSSA, was established as a journal in 1972 to provide an outlet for technical reports and reviews that are concerned with the protection and improvement of environmental quality in natural and agricultural ecosystems. It focuses on environmental quality work, to provide recognition for scientists whose research is directed toward this area of investigation and to make it possible for scientists in other disciplines to locate and recognize contributions to environmental quality.

The journal went from quarterly to bimonthly publication in 1994. In 2005 (Vol. 34) all manuscripts were submitted electronically, with 41% of the articles submitted originating from outside the United States. The publication fee was changed to $650 per manuscript for all articles.

Subject matter categories were implemented in 1991 with Vol. 20 of JEQ. Manuscript Tracker was implemented for submitting and reviewing manuscripts first in JEQ in 2001. In 2002 JEQ began publication of the online version with HighWire Press. Since 2002, JEQ is published in printed copy, CD, and online versions. A three-CD set for JEQ (Vol. 1–30, 1972–2001) became available for purchase in 2004, and these back issues also are being made available online in 2008.

Journal of Agronomic Education/ Journal of Natural Resources and Life Sciences Education

The ASA initiated *The Journal of Agronomic Education* (JAE) with other cooperating societies in 1972, with one issue per year, for the purpose of providing an opportunity for publishing education articles by instructors in extension and resident education. In 1984, JAE was published twice each year. In 1992 the name was changed to *Journal of Natural Resources and Life Sciences Education* (JNRLSE) with an expanded scope and coverage. The total number of pages increased from 120 to 200 (1999–2004), with a reduced number in 2005. The publication fee was $350 per manuscript for all articles and $150 for web lessons. In 1998, JNRLSE was the first journal with an online version, with articles posted as they were ready to publish (continuous publication). Electronic submission and review for JNRLSE began in 2004.

Journal of Production Agriculture (JPA)

The JPA was a technical peer-reviewed journal established by ASA, CSSA, and SSSA and was published quarterly beginning in 1988. Before that, production agriculture articles had been published in AJ. The JPA was created primarily to transmit production-oriented information in an understandable and useful manner. It was intended for a wide range of professionals, including scientists, consultants, journalists, farm advisers, farm managers, extension specialists, teachers, and college students. Because of low circulation and mounting costs, the journal was terminated in 1999. After 1999, production articles formerly published in JPA were published in AJ.

Plant Management Network Journals

The Societies initiated *Crop Management* as an online journal in 2002 published by the American Phytopathological Society (APS). *Forage and Grazinglands* and *Applied Turfgrass Science*, sponsored by CSSA and ASA and published by APS began in 2003 and 2004, respectively. They can be accessed online at www.plantmanagementnetwork.org, owned by APS. These journals are targeted to practitioners and feature continuous publication and an abundance of graphs and photos.

Crops and Soils Magazine

Crops and Soils magazine was first published in 1948 as *What's New in Crops and Soils* (1948–1958). It was renamed *Crops and Soils Magazine* (from 1959 to October 1987) with subscribers reaching 23,114 in 1977. Since its inception, the Society had provided support beyond the publication's earnings even though promotion campaigns helped reduce the deficits. The Budget and Finance committee decided the publication was losing too much money and it ceased publication in 1987 with the 39th volume.

In 2006, ASA decided to start publishing a magazine aimed at Certified Professionals, and *Crops and Soils* was reborn. The first issue of the new series (Vol. 40, No. 1) was released on March 15, 2007.

CSA News

The Society newsletter/magazine, formerly known as *Agronomy News*, underwent several redesigns during the past 25 years. The name was changed to *Crop Science, Soil Science, and Agronomy News* and in 2000 to *CSA News*. A major change was made with the January 2006 issue. It is now in a magazine format, is printed on glossy paper, and includes color. *Agronomy News* has always been a source for current communications about science and member news. Now it contains many other features, including technical articles of broad appeal.

Newsflash

A free, electronic newsletter for members-only, called the News Flash, has been sent via email bi-weekly to ASA-CSSA-SSSA members as a member benefit since March 2002. Written by staff, it includes information and announcements that are new, deadline-driven, or have a call-to-action relating to the products, services, and activities of the Societies. Since November 2005, the Societies have published a graphically enhanced News Flash (adding photos, color, more bullets, etc.).

Other Publications

In addition to the journals, ASA, CSSA, and SSSA publish several additional publications. The ASA Special Publication Series, often published in cooperation with the other two Societies, began in 1963. To date there are 66 titles in this series. ASA also has co-published titles in both the CSSA and SSSA Special Publication Series.

The Agronomy Monographs series began in 1949 and has continued throughout the last 30 years. There are a total of 49 titles in the series, with many volumes being reissued as second and even third editions. ASA published numerous other books, some in the Foundations for Modern Crop Science series, and many others independently or cooperation with other societies and groups summarizing symposia or technical subject areas. A complete list of ASA publications can

be found on the accompanying CD. Publishing has changed significantly in the last 30 years and undoubtedly will continue to change in the future. Changes include both obvious changes, such as cover redesigns and changes behind the scenes, with the use of new technologies such as computer editing and desktop publishing. A challenge to ASA will be to continue to adapt and meet the needs of members and readers.

OUTREACH ACTIVITIES

Collaboration

The Societies have participated in a number of consortia that have emerged to meet common goals for public awareness, promotion of research, and efficiency of operations. Examples of such consortia have included the following: American Institute of Biological Sciences, American Association for the Advancement of Science, Council for Agriculture and Science Technology (CAST), Coalition for Funding Agricultural Research Missions (CoFARM), Council of Scientific Society Presidents (CSSP), Coalition for National Science Foundation (CNSF), and the Renewal Natural Resources Foundation (RNRF).

The three Societies, ASA, CSSA, and SSSA, became charter members of CAST in 1978. The Societies have co-sponsored many special activities. For example, in 1988, the World Bank and the Societies sponsored a workshop on Agricultural Sustainability. ASA members have periodically served as North American representatives to the continuing committee of the International Grassland Congress (IGC). In 2001, ASA initiated active participation with the IGC as a sponsoring society of the Canadian Congress. In 2005, the IGC requested that ASA develop and manage its permanent website. ASA is a sponsoring society for the joint meeting of the International Rangeland Congress and the IGC to be held in China in 2008. Additional information on international cooperation is included in the chapter on international activities (Ryan, 2007, this publication).

In the late1990s, ASA was active in increasing further collaboration with the European Society of Agronomy (ESA). A memorandum of understanding (MOU) was developed with the ESA in 2004, establishing complimentary registration for up to three individuals to the others annual meeting and linking to the others web site on their home page. An ASA–CSSA–SSSA and Canadian Society of Agronomy MOU was approved in 2007 to increase the interaction with that society. The opportunity exists to develop similar MOUs with other cooperating societies. The new structure will facilitate future discussions with other organizations that may have an interest in affiliation.

Science Policy Programs

Issues facing national policymakers and decision makers often contain elements that require scientific or technical expertise. Members of Congress, their staffs, and Administration personnel often do not have the technical expertise needed. It is impossible for any national leaders to have a technical background in each of the many diverse issue areas that they must address. Thus, it is important that the technical and scientific communities provide the information that policymakers need in a form that can be understood at the time it is needed. The Societies, responding to the need for engaging scientists and professionals in national policy discussions, have taken a lead role in exploring various means for providing science-based professional expertise into federal public policy discussions emphasizing agriculture, natural resources, environmental sciences, and education.

Congressional Science Fellowship Program

The purpose of the Congressional Science Fellowship Program is to make practical contributions to the more effective use of science and technical knowledge in government, to demonstrate the value of such science–government interaction, to inform the scientific communities about public policy and the legislative process, and to provide a unique public policy learning experience. The criteria for selection of Congressional Science Fellows are aimed at highly qualified and interested persons in early or mid-career. As of 2005, applicants are required to have their Ph.D. at the time they start the fellowship. Candidates must have exceptional competence in some area of the agronomic, crop, soil, or related fields of science and education; be cognizant of a broad range of matters outside the Fellow's particular expertise; and have a strong interest in working on a range of public policy issues. The Fellows are free agents who become involved in either a Senator or Representative's office or on a Congressional Committee, but are not asked to carry an agenda or be an advocate for the Societies. One of the responsibilities of the Congressional Science Fellow is to publish a monthly article on their activities in *CSA News*.

The initial program was started in 1986 with the appointment of Jonathan D. Haskett, a student with a Masters degree in soil science from the University of Minnesota. Jonathan served from February 1986 through January 1987 in the office of Senator Quenton Burdick, Chair, Senate Committee on Environment and Public Works. The Societies formally established the program in 1987, and after interviewing a number of candidates, selected Dr. Terry L. Nipp, a graduate of Oklahoma State University. Terry served from May 1987 through May 1988 with Representative George E. Brown Jr. on the House Committee on Department Operations, Research, and Foreign Agriculture. Since 1987, a committee representing the Societies sponsoring the program has formally screened the applicants. The Fellowship Program has been affiliated with the American Association for the Advancement of Science (AAAS) since 1998, when our Congressional Science Fellows became involved in the AAAS formal orientation program in Washington, DC. From 1989 to 1998, the sponsoring societies included the national and regional Weed Science Societies as well as ASA, CSSA, and SSSA. As of 2007, 29 Congressional Science Fellows have served the Societies. These alumni are in various university, industry, and governmental positions, including many in influential positions, both in Congress and the Administration, involving decision making that directly influences policy issues and funding for our sciences.

Science Policy Office

As a means of enhancing the Washington presence of the Societies, a Washington liaison position was created to address those issues and activities on behalf of the Societies. Initially the liaison representative was established through a contractual agreement with AESOP Enterprises, an independent firm that provided consulting and liaison representation services. This agreement was active from 1995 until 1999 when the Office of Science Policy was established through the employment of a director of science policy (DSP). Since the inception of the Societies Science Policy Office in 1999, Dr. Karl M. Glasener has served as the DSP. The Science Policy Office works with Executive (federal agencies and White House offices) and Congressional personnel offices and committees to promote legislation, rules, regulations, and funding supportive of the agronomic, crop, and soil sciences. On behalf of the Societies, the Science Policy Office submits testimony to Congress stressing funding priorities and the DSP meets with both Congressional and Executive offices to educate policymakers and advocate for our sciences. These activities have been critical in providing justification for continued support and enhanced federal funding for science.

The major purposes of the DSP were to provide Society officers, staff, and program leaders with continuous and timely information on the status and impact of science policy developments in Washington, DC; recommendations of science policy initiatives worthy of attention and support by the Tri-Societies; and guidance on the appropriate role of the Societies in science policy leadership, support, liaison, and networking. The director serves as a scientific resource for policymakers in the general areas of the agronomic, crop, and soil sciences. In 2004, ASA developed a Rapid Response Team whereby the DSP can obtain names of scientific experts in the ASA divisions to provide technical information or for nomination to federal advisory committees, boards, or councils. This is an effective method to provide scientific information to policymakers in a timely manner. The Director also has been active in various consortia such as CoFARM, National Coalition for Agricultural Research (NCFAR), and CNSF. He chaired the CoFARM group, which helped draw large numbers of scientific societies together in interacting with policymakers, with the goal of achieving significant increases in funding for research programs sponsored by USDA.

(Left) Karl Glasener heads the Societies' Science Policy Office in Washington, DC. These efforts constitute an important link between the federal government and the agricultural, natural resources, and environmental communities. (Right) CEO Ellen Bergfeld meets with Representative Tammy Baldwin, one example of the many meetings between Society leadership and policy makers over the years.

A Science Policy Internship Program was started by the Societies in 2001. The Science Policy Internship provides qualified individuals in the agronomic, crop, soil sciences, and related disciplines with hands-on learning experience in science policy. The internship offers flexible start and end dates and can be for three to six months. The intern works closely with the DSP monitoring and analyzing agricultural, natural resources, and environmental legislation; attending and participating in Congressional and Agency hearings and briefings; assisting in the preparation of position papers; directing the Societies' grassroots advocacy activities; and communicating with members of the Societies. Sara Hayes, an M.S. degree candidate in soil science at North Carolina State University was the first intern, serving from July to September 2002. Through 2007, seven individuals will have served as Science Policy Interns (see the list on the accompanying CD).

Marketing Communications for the Future

As agronomy, crop science, and soil science continued to evolve, marketing communications are becoming an increasingly important tool for recruiting new members to the Societies, educating students about career opportunities, informing the public about the importance of the Societies, and ensuring that policymakers support the work of the persons involved in the sciences. The tools used to market and communicate have changed significantly in the last 30 years. Today web-based technologies are used in conjunction with more traditional mediums such as print or direct mail and are often the most effective ways to reach international audiences. As high-speed internet access becomes more widespread and inexpensive, the Societies will use email, podcasts, RSS (really simple syndication) feeds, and other technologies still on the horizon. Although the tools may change, marketing communications will fundamentally remain about telling the ASA story.

Information Technology

The pace of change over the past 100 years is most evident in the use of technology. Personal computers at everyone's desk are the norm in 2007, but that wasn't the case prior to the late 1990s. Technology has helped staff track member/customer data to better understand and serve our customer/members. The Societies Office first began using computers in 1980. These computers used the WANG operating system and were largely used to manage membership and subscription information up until 1984. Then, the Societies Office converted to the IBM System 36 Platform from 1984 to 1999. The IBM AS400 database helped employees manage membership and subscriptions from 1999 to 2006. In 2006, the Societies Office migrated to Protech, a Microsoft-based system that uses the latest in CRM technology. In addition, technology, specifically the Internet, has enhanced the way members submit abstracts (online-only beginning in 2000), submit awards (online-only beginning in 2002), register for meetings, order books, renew their membership, and email staff. Online delivery of the journals in 2007 may be regarded as conforming to the best current standards in online scientific publishing, ensuring our content is well archived and widely read. We also have three Society websites, which were launched in the year 1988: www.agronomy.org, www.crops.org, and www.soils.org. In March 2007, 1,726,657 hits were recorded to the ASA website. In 2006, the websites added a search engine using the latest in Google technology. In addition, employee email was started in 1988.

Career Placement Center

The Career Placement Center, formerly called Placement Service, was established in the mid-1960s by Matthias Stelly, Executive Vice President, and was a new membership service that acted as a clearinghouse for resumes and personnel listings. The service also included interview scheduling between applicants and employers during the annual meetings.

The Career Placement Center brings together hundreds of employers and applicants every year.

In the mid-1980s, the Headquarters staff prepared resumes from a form completed by the applicant, and a supply of resumes was printed and sent to applicants for their use, with some retained to make referrals to employers. Up to the late 1990s resume preparation was done using a custom computer program that worked off of the membership database. From 2000 to the present, resume submission is done by the applicants via a web-based system, and employers can obtain resumes through a database search function. The Career Placement Center was completely redesigned with added services in 2007. In 1988, a new computerized scheduling system was implemented and used until 2000, when a web-based program was created and used until 2005. In 2006, the scheduling system was redesigned and implemented during the 2006 Indianapolis meetings.

The Placement Service was overseen by the ACS 527 Placement Committee. In 1987 this committee merged with the Career Promotion Committee, and the name changed to the Placement and Career Promotion Committee. As a result, the name of the service changed to the Career Placement Service. In 2003, the committee merged again and became part of the ACS 537 Membership Committee, and the name changed to the Career Placement Center. In 1975, the 1964 brochure "Agronomy and You" was revised and reissued as "Careers in Agronomy, Crop Science, & Soil Science." It was revised again in 1986, 1989, and 1995, when the name changed to "Careers in Agronomy, Crop, Soils, & Environmental Sciences." In 2000, the brochure was discontinued and replaced by a soils brochure, "Soils Sustain Life," and a crops brochure, "Grow Your Future."

Certification Programs in ASA

For many years the Societies have worked toward the best means for disseminating scientific information to the various potential audiences. In 1969, a special committee was formed to begin work on developing standards for certification in agronomy. Several years were required for the procedures to be finalized. The operation of the American Registry of Certified Professionals in Agronomy, Crops and Soils (ARCPACS) began in 1977. The first structure was organized as an affiliated corporation to ASA. It had three sub-boards, one each for agronomy, crops, and soils, and one overall advisory board, called ARCPACS, with its own director and support staff shared from ASA. This early structure lasted until 1983 when ARCPACS as an affiliated corporation was ended. It then became a program within ASA due to financial difficulties. This new structure allowed the program to grow as a membership service of ASA into the early 1990s. In 1991, ASA approved certification categories of weed scientist, plant pathologist, and horticulturalist. Later that same year, ASA approved in concept the creation of the certification category for certified crop adviser. ARCPACS remained a membership service of ASA and grew to seven certifying sub-boards: agronomy, crops, soils, horticulture, plant pathology, weed science, and crop adviser. The name was changed in 1993 to better depict the addition of the other sub-boards outside of agronomy, crops, and soils. The new name was ARCPACS: A Federation of Certifying Boards for Agriculture, Biology, and Earth and Environmental Sciences. The old acronym remained in use along with the subtitle. ARCPACS grew to a high of about 2700 certifications by the early 1990s across the six certifications.

The Certified Crop Adviser Program (CCA) operated separately from ARCPACS; thus, those numbers are not included in the AR-

Today more than 15,000 people certified in ASA programs bring the latest science to the fields.

CPACS totals. The CCA program began operation with the first exams being offered in 1993 and continues through today with 13,568 CCAs throughout the United States and Canada. CCA started out as only a U.S. program but quickly grew to add Canadian provinces in 1997.

ARCPACS and CCA worked together but were structured differently. The six ARCPACS sub-boards were national in scope, while CCA started out as a state-based program with one national policy board. Each state or province that entered the program had to organize a state or provincial certifying board.

In 2002, the horticulture certification was given to the American Horticultural Society in hopes that it would gain more recognition being connected to its scientific society. By 2003, ARCPACS had declined to 2,030 certifications, with the majority of them in either the soils certifications (1200) or agronomy (700). It was decided to have SSSA perform the administrative functions for the Certified Professional Soil Scientist and Classifier programs, with ASA maintaining the Certified Professional Agronomist and Certified Crop Adviser programs. Each related society to the crops, plant pathology and weed scientist programs was asked if they wanted to assume the administrative functions. Due to the very low numbers in each of the programs, they were discontinued. In 2007, ASA has two certification programs: International Certified Crop Adviser (ICCA) with 13,568 certified and Certified Professional Agronomist (CPAg) with 658 certified. The Soil Science Society of America has the Certified Professional Soil Scientist (CPSS) with 955 certified and Certified Professional Soil Classifier (CPSC) with 75 certified.

AGRONOMIC SCIENCE FOUNDATION: FORTY YEARS OF SERVICE TO AGRONOMY, CROP, AND SOIL SCIENCES

Organization and Governance

The Agronomic Science Foundation (ASF) was founded by seven ASA, CSSA, and SSSA leaders in 1967 in response to a 1964 recommendation of the Society's Organization and Policy Committee. The concept of a foundation was advanced with the objective that resources (funds) could be developed to support education and research programs of the three Societies. At a meeting in Chicago on November 11, 1968, Rod Bertramson declared, "We are in business." After approval of ASF as a nonprofit, charitable organization under Section 501(c)(3) of the U.S. Internal Revenue Code of 1954, it was chartered in the State of Wisconsin. The Articles of Incorporation and Bylaws were published in Agronomy Journal 60:133–135 in 1968. The American Society of Agronomy has served as the management office of ASF since its inception.

ASF is governed by a Board of Trustees with members initially appointed by the President of ASA from members of one or more of the three Societies and nominated by the Executive Committees of the three Societies. The board was composed of three representatives from each Society, three members-at-large appointed by the ASF board, the chief executive office of the Societies, and two additional members appointed by the ASF Board. The member appointment process was changed when the ASF bylaws were revised in 2003. All members are now elected by the board from recommendations solicited from Society presidents. In the absence of recommendations from the Societies, the board prepares a roster of nominees. Terms of service are normally for three years with possible re-election for a second three-year term by the board of directors. Board members serve without compensation.

Since its inception, 77 individuals have served on the board. From 1967 to 2000, the board elected a president and vice-president. In subsequent years the titles were changed to chair and vice-chair. With the change in bylaws in 2003, the executive vice-president of the three Societies was retained as permanent ex officio, nonvoting member, and secretary-treasurer. The first president of ASF was H.F. Robinson from North Carolina. (See roster of ASF officers and board members on the CD.)

Mission and Purposes

The purposes of ASF, as identified in the Articles of Incorporation, are "…to organize this foundation solely for scientific, educational, and charitable purposes in the broad interest of agronomy, crop science, soil science and other related disciplines, and in the furtherance of these ends it shall administer such money and property as it may receive, but the ASF shall engage in no business activity except the investment and reinvestment of its assets."

The mission of ASF has evolved over time. In 1992, the "primary mission of the ASF is to secure financial support for research and educational programs of ASA, CSSA, SSSA. The programs must be scientifically sound technically, economically feasible, as well as protective of the environment." The current mission statement is "to provide leadership and financial resources to further the role of the agronomic, crop, and soil sciences in global crop production, and to promote human welfare within a sustainable environment."

Operating Procedures

Beginning in 2000, attention was given to security of donor privacy and protection of ASF assets, and protection of its 501(c)(3) status with the IRS. The Articles of Incorporation and Bylaws revised in 2003 specified this action. Segregation of ASF funds from the pooled assets of the three Societies was effected to deter outside access to ASF funds and to manage the portfolio.

Audit reviews are required to obtain transparency in order to retain donor satisfaction. During the past two years, all operational procedures have been reviewed and are in continual upgrading mode. An ASF Policy and Procedures Manual is in revision as a public document.

Functions of ASF are facilitated by ad hoc and standing committees composed of board members, including a Development Advisory Committee, Budget and Finance Committee, Audit Committee, Investments Committee, and Nominations Committee. The committees are assisted by headquarters staff.

A review of operational costs from 1989 to 2006 illustrates three phases during the 40-year history of ASF. Prior to 1989, very little investment in ASF activities was incurred. In the eight years since 1989, annual operational costs ranged from $2,562 to $16,970, with a mean of $12,550 per year; in the next five years, 1997–2001, the costs ranged from $30,622 to $61,845 for a mean of $40,706. In 2001–2006, the costs ranged from $50,906 to $133,791 for a mean of $111,818. These expenses during the past 18 years reflect increasing fund-raising activities and specific fund developments. In 1997, upon the request of the ASF Board of Trustees, the first Director of Development, John Kruse, was employed to concentrate on fund-raising as an ASF service to membership in the three societies and had considerable success. From 1997 to 2001, ASF incurred personnel costs. Beginning in 2002, a three-year rolling grant commitment of $100,000 per year was formalized by the three societies to support salary and some expenses for ASF development of programs and funds by ASF. Society funding of ASF for development activities was shifted directly to each Society in 2006 when more detailed accounting of costs was implemented.

Since 2006, ASF has used its own sources of operating funds rather than direct grants form the Societies. During 2006 and 2007, the operating costs have been met from undesignated donation funds (Priority Fund). Since this is unsustainable and beyond the intentions of donors, reviews are underway to explore the generation of operating funds for the Foundation. The operations of the three Societies and the Foundation are operating on a cost-for-services basis, thus operating on a more direct business model.

The first Director of Development hired by ASA for ASF, John Kruse, resigned after about one year. The second Director of Development, Valerie Breunig, was hired in 1999 and served until 2004. In 2005, two development officers, Bonnie Leuck and Paul Kamps, were appointed by ASA with funds from the three Societies for service to all three Societies; service to ASF is provided on a time-recharge basis. The assets of the foundation and the number of endowed programs grew greatly during the past eight years.

Fund Development

The mission of ASF has been to accumulate funds to support research and education activities. ASF implicitly serves to advance programs for the benefit of the members of the three Societies. To date there were no benefactors that were not approved by at least one of the Societies. Donation funds were solicited by CSSA and ASF for inauguration of the International Crop Science Congress in the 1990s and by SSSA and ASF for use by the Smithsonian Institution to develop a soils display as a project of SSSA from 2003 to the present. ASF has funded scholarships that are awarded to nonmember students.

Most of the funds are designated to support the legacies of individuals or families in the form of scholarships, lectureships, and/or conferences. As of 2006, there were 40 such funds. Most of the funds are endowed at a level of $25,000 or more. From a very modest beginning of $26 in 1969, the funds have grown to $2.7 million in 2006 from 7,756 donations (1988–2006).

At the end of 2006, ASF's assets in cash and investments were a sizeable $3,075,457. Permanently designated funds in endowed and nonendowed funds were $1,209,697, or 39.3% of the total assets. Of the remaining $1,865,760, all but $125,660 were assigned to specific funds and their use restricted to supporting those designated activities. Thus, ASF's funds for operational expenses and new programs were only 4.1% of its total assets. With the changes since 2005, undesignated funds are being drawn down for operations and minimally for programs, a clear indicator of the unsustainability of ASF under current operational procedures.

Program Areas and Benefits to the Three Societies

The programmatic emphases of ASF are guided by the Society strategic plans, especially since 1966, and during the period 1991–1996 ASF annual reports identified five focus areas for fund development:

Human resource development
Environmental and natural resources stewardship
Science, technology and public policy
Current issues, forums and workshops
Professionalism and recognition of quality and excellence

These focus areas have been generally retained and provide guidance to the Societies, trustees, and donors. A rich roster of activities serving the Societies' membership and students can be accessed in the ASF website: https://www.a-s-f.org.

ASF: Looking Forward to the Next 40 Years

ASF will continue to serve as a recipient and curator of funds dedicated to the legacies of outstanding scientists consistent with strategic plans of the three Societies such as lectureships, conference, and scholarships. The Foundation will participate in special funding activities as it has in the past. CSSA, with ASF, has embarked on a vigorous campaign to support its Golden Opportunity Scholars Program for undergraduates. SSSA and ASF are nearing the end of a successful campaign to fund a major soils exhibit at the Smithsonian Institution's National Museum of Natural History in Washington DC, which will run from July 2008 to January 2010.

On the operations and management side, ASF is reviewing means for providing sources of operating funds by collecting management fees from donations and investment income. Clarification of relationships with the

Members of the 2007 ASF Board of Trustees: (from left) Ellen Bergfeld, S.K. De Datta, Mary Collins, Cal Qualset, Lee Sommers, Jim Coors, Vivien Allen, Ron Cantrell, Wayne Keim, George Ham, and Rick McConnell. Not pictured are Don Plucknett, Tom Simms, and Jim Watson.

The International Annual Meetings of the American Society of Agronomy, Crop Science Society of America, and Soil Science Society of America today attract more than 4000 attendees from around the world. The meetings bring together people representing academia, government and private industry, including a large contingent of undergraduate and graduate students

three Societies to establish most effective means for fund enhancement to facilitate Society programming goals will be studied and clarified during this centennial year for ASA and the fortieth year of ASF.

ANNUAL MEETINGS

The American Society of Agronomy has held an annual meeting since the organizing meeting in 1908, with the exception of 1944 during WW II. A central theme has been chosen for the annual meetings since 1962. Revisiting those themes provides information on the changes in focus through the years and a paradigm of the growth in influence provided by the Societies (see the list of meetings and themes since 1962 on the accompanying CD). The Societies have evolved from a focus primarily directed at production agriculture to a wider scope that includes consideration of urban clientele, environment and sustainability, energy possibilities, biotechnology, science policy for working with governmental decision makers and agencies, and ultimately serving as a global forum in which to exercise activities, provide information, and exert influence. Beginning in 2005 each Society could select a unique theme.

IMPACT STATEMENT

The American Society of Agronomy's objective is to enhance the science and profession of agronomy. As with any membership-based organization, it depends on its members for any achievements made to meet that objective. Its achievements through the first 100 years have been outstanding. Membership roles of the Society boasts recipients of the Nobel Peace Prize, World Food Prize, members in the Academy of Sciences, pioneers and leaders in the Green—now Evergreen—Revolution that provided and will provide resources to feed starving peoples of the world, developers of hybrid crops that achieved major yield increases, and scientists at the forefront of the work in biogenetics and genetic engineering that will one day solve complex problems in energy, health, and hunger, while developing new practices that enhance and preserve the environment.

The American Society of Agronomy will be the members' and society's indispensable resource for leading-edge education, knowledge, and networking. It will be recognized as the powerful advocate and voice for advancing the science of agronomy to proactively address emerging global and social issues. The integrated science of agronomy will be recognized as the source of science-based knowledge that improves and integrates the management of soils, crops, and the environment. The organization itself has worked and is tirelessly working to provide the latest information on scientific research; to nurture state, national, and international relations; and to promote professionalism in all aspects of the conduct of agronomic research and its dissemination from the research plot or laboratory to the ultimate clientele in the field. In the past 100 years, ASA and its members have proven to be resilient and progressive in foresight. As ASA looks to the future, it will need to be in the forefront of important dialogs among scientists and policymakers to solve society's complex problems.

REFERENCES

Laude, H.H. 1962. History of the American Society of Agronomy, First fifty years—1907–1957. Agron. J. 54:57–69.

Lyon, T.L. 1933. History of the organization of the American Society of Agronomy. J. Am. Soc. Agron. 25:1–9.

Milford, M.H., and D.T. Smith. 2007. Development and evolution of resident education in the American Society of Agronomy. p. 69–78. *In* L.E. Moser (ed.) The American Society of Agronomy: 100 years of history. ASA, Madison, WI.

Payne, W.A., and J. Ryan. 2007. The international dimension of the American Society of Agronomy: Historical perspective, issues, and challenges. p. 89–98. *In* L.E. Moser (ed.) The American Society of Agronomy: 100 years of history. ASA, Madison, WI.

Pearson, C.H., S.M. Ernst, K.A. Barbarick, J.L. Hatfield, G.A. Peterson, and D.R. Buxton. 2007. *Agronomy Journal* turns one hundred. p. 59–68. *In* L.E. Moser (ed.) The American Society of Agronomy: 100 years of history. ASA, Madison, WI.

Smith, D.C. 1980. Development of the American Society of Agronomy, 1958–1977. Agron. J. 72:227–240.

Agronomy Journal

PROCEEDINGS

OF THE

American Society of Agronomy

VOLUME I

1907, 1908, 1909

PUBLISHED BY THE SOCIETY
1910

Agronomy Journal Turns One Hundred

During 2008 we celebrate the centennial anniversary of *Agronomy Journal*. Many people have certainly been influenced in some way by the science published during the 100-year existence of the journal. From Volume 1 up through Volume 98 (2006) there have been more than 30,290 authors who published 15,232 articles totaling 89,056 pages. More than 2545 editors were required to review and edit the papers published in *Agronomy Journal*, in addition to the manuscripts submitted but not published. As a current snapshot of *Agronomy Journal*, we published 60% of the manuscripts submitted in 2005. In both 2003 and 2004, we accepted 55% of the manuscripts submitted. In a comparison of 48 peer journals in 2005, the impact factor of *Agronomy Journal* ranked 12th at 1.473 and the total citations for the journal ranked fourth at 6723. Commentaries on the early history of *Agronomy Journal* have been previously published. In our article, we focus on the journal's history during the past 25 years. We fully expect that the future of *Agronomy Journal* will be even more exciting, rewarding, challenging, and valued as the past 100 years. We eagerly look forward to the next 100 years of *Agronomy Journal*.

Calvin H. Pearson
Colorado State University, Fort Collins

Susan M. Ernst
ASA–CSSA–SSSA Office, Madison, Wisconsin

Ken A. Barbarick
Colorado State University, Fort Collins

Jerry L. Hatfield
USDA-ARS National Soil Tilth Laboratory, Ames, Iowa

Gary A. Peterson
Colorado State University, Fort Collins

Dwayne R. Buxton
USDA-ARS, Albany, California

During 2008 we will publish volume 100 of *Agronomy Journal*—marking the centennial anniversary of the journal. Also, this first issue of 2008 debuts a redesign of the journal. The publishing industry is undergoing rapid and substantial changes, much of which is to our advantage. This gave us an opportunity to redesign *Agronomy Journal* at a low cost and we chose to schedule it as part of the centennial anniversary.

The Volume 100 redesign included consideration of the following: journal cover redesign (layout and format orientation, color choices and combinations, fonts and sizes), article title page layout, updated font combinations and sizes, ragged right versus full justification, two-color, two versus three columns, footers, headers, page number placement, sidebars, pull quotes, increased use of photographs, shading, and shadows.

The first four volumes (1908–1912) of *Agronomy Journal* were titled the *Proceedings of the American Society of Agronomy* and Volumes 5 through 40 (1913–1948) were titled the *Journal of the American Society of Agronomy*. Following a vote by the membership, the name was changed to *Agronomy Journal* in January 1949. At the same time, the format was changed from 6 by 9 inches to a trimmed size of 8.5 by 11 inches (ASA, 1948b). *Agronomy Journal* has retained its current name since 1949.

During the 100 years of *Agronomy Journal*, there have been ongoing changes to the journal, including six major design changes. These previous design changes first appeared in Volume 5 (1912), Volume 39 (1947), Volume 41 (1949), Volume 75 (1983), Volume 91 (1999), and Volume 100 (2008).

With the journal reaching such a significant milestone we are presented with a unique opportunity to celebrate, reflect, and anticipate the

A preprint from
Agronomy Journal, Vol. 100, January–February 2008

C.H. Pearson, Colorado State University, Fort Collins, CO 80523; (email: calvin.pearson@colostate.edu).

Left: The first page of the the first volume, from the Society archives.

Above: The Agronomy Journal *centennial logo.*

Section page: Covers through the years, on a backdrop of Carleton's words.

future. Many people, both in and out of agriculture, too numerous and too varied to mention, have been influenced in some way by the science contained in the 100 volumes of *Agronomy Journal.*

In pondering 100 years of *Agronomy Journal,* various questions come to mind. While these questions may have no definitive answers they are nonetheless questions worthy of reflection. Our intent with these questions is to encourage the reader to contemplate the contributions, challenges, and value of this publication during the past 100 years and on into the future. The questions are not presented in any particular order and many more questions could be added to this short list.

What papers, published in *Agronomy Journal,* have had the greatest positive impact on society?

How has the way we conduct agronomic research changed during the last 100 years?

How have the topics published in Agronomy Journal changed in the past 100 years?

Have we met important needs of society through the research findings published in the journal during the past 100 years?

Is *Agronomy Journal* well positioned to meet the needs of our society and our profession for the future?

HISTORICAL DATA FOR *AGRONOMY JOURNAL*

From Volume 1 up through Volume 98 (2006) there have been more than 30,290 authors who published 15,232 articles, totaling 89,056 pages (Fig. 1). More than 2545 editors (the number of editors was counted by adding the number of people serving on the Editorial Board for each year) were needed to review and edit the papers published in *Agronomy Journal* (Fig. 2), in addition to the manuscripts submitted but not published.

The following facts about *Agronomy Journal* may interest the reader.

- The largest volume of *Agronomy Journal* was published in 2004 (Fig. 1C). Volume 96 contains 1828 pages and consists of 208 articles.
- The single largest issue of *Agronomy Journal* published from the period 1908–2006 was the May-June 2006 issue. This issue is a hefty 456 pages (460 pages with front matter).
- Volume 61 (1969) contains the most articles at 302 (Fig. 1B) and has 1017 pages. Volume 65 (1973) has the second most articles at 299 and total pages are 1062.
- The most authors in a single year (794) was in 2006 (Vol. 98) (Fig. 1A). If someone authored more than one paper, they were counted each time they were listed as an author on an article.
- For years where data are available, the largest number of reviewers needed to publish a volume of *Agronomy Journal* was 654, which was Vol. 92 (2000) (Fig. 3). The number of reviewers needed to review papers submitted to the journal has come close to or exceeded 600 in many recent years.
- The number of ASA members who subscribed to *Agronomy Journal* peaked in 1985 at 7742 while the number of nonmembers (e.g., libraries, companies, organizations, individuals) peaked in 1976 at 2151 (Fig. 4). The number of members who subscribe to *Agronomy Journal* has decreased every year from 1985 through 2005, and for

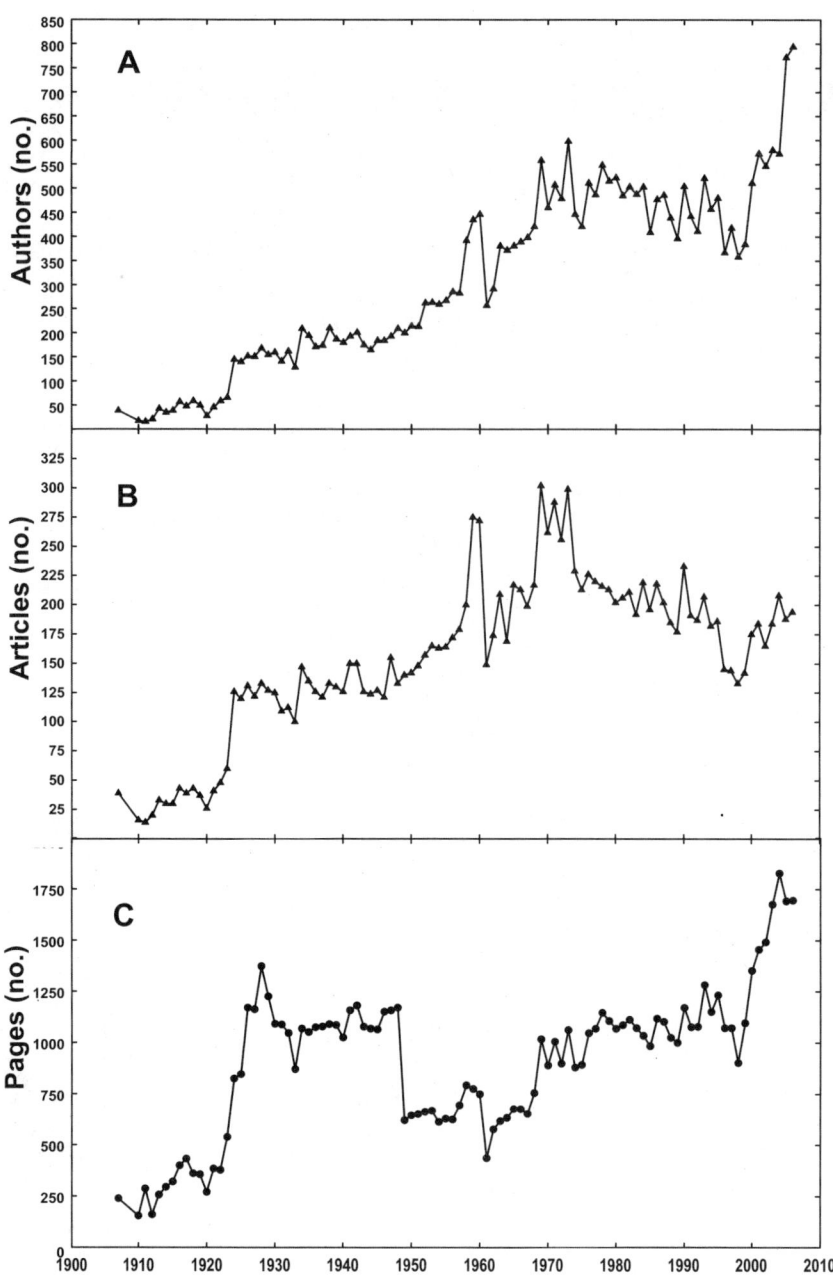

Fig. 1. Numbers of (A) authors, (B) articles, and (C) pages included in each volume of Agronomy Journal *from 1908 to 2006. The same author may have been counted more than once if the author had more than one paper in a given year.*

nonmembers this decrease has also continued nearly every year since peaking in 1976. In 2006, the total number of subscriptions was down from 2005 by 205 due to a drop in nonmember subscriptions; however, after experiencing decreases for 20 years, member subscriptions increased by 434 in 2006. Such a sustained decline in both member and nonmember subscribers to *Agronomy Journal* has created ongoing concern over the long-term health of this publication.

During the 100 years of *Agronomy Journal* there have been 13 Editors (Fig. 5). Their affiliations and service periods are presented in Table 1. Both J.D. Luckett and Matthias Stelly each served as Editor for 21 years. Dr. Buxton was appointed as both Editor and Editor-in-Chief of the American Society of Agronomy. After 2 years, the Executive Board of the Society, with concurrence from Dr. Buxton, decided to separate the two positions.

In the early years of *Agronomy Journal*, only three to five people comprised the Editorial Board. In recent years, the number of people serving on the Editorial Board has reached nearly 75 (Fig. 2).

CHANGES IN *AGRONOMY JOURNAL* IN RECENT YEARS

Commentaries that include historical information about *Agronomy Journal* have been previously published (Lyon, 1933; Throckmorton, 1941; Laude et al., 1962; Smith, 1980; ASA, 1983b; Fuccillo, 1983). Lyon (1933) provided an early history of the journal, dating from its beginning up to 1930. The Diamond Jubilee of the American Society of Agronomy was celebrated in 1982 and Volume 75 was published in 1983. The historical highlights published in *Agronomy Journal* (ASA, 1983b) focuses mainly on the society but also contain some historical information related directly to the journal. Laude et al. (1962) and Fuccillo (1983) have provided detailed histories of the first 75 years of Agronomy Journal. These articles are invaluable to those who are interested in the evolution of *Agronomy Journal*. The major objective of our article is to document the journal's history during the past 25 years.

1982–1983

Dwayne R. Buxton, currently the oldest living former Editor of *Agronomy Journal*, served in this position for 2 years while at the same time serving as Editor-in-Chief of ASA. It was during his tenure that the two positions were separated (ASA, 1983a). During Dr. Buxton's relatively short time as Editor several significant changes occurred. The Editor's position was moved from a paid position at Headquarters to a nonpaid scientist working at a location away from Headquarters. This occurred in 1983, when his two appointments were separated (ASA, 1984a). Subsequently, Dr. Gary A. Peterson was appointed as Editor of the journal and Dr. Buxton continued to serve as Editor-in-Chief for four more years.

The SI system (Système International d'Unités) of reporting units was introduced but not without encountering some resistance and controversy from peer scientists (ASA, 1982). The handbook of instructions for authors was revised in 1982 and subsequently published (ASA, 1983a, 1984b, 1985) along with a handbook to help editors (ASA, 1986, 1987) to be more effective in handling manuscripts and working with authors.

1984–1989

Before 1984, all manuscript submissions were sent to Headquarters, which was appropriate when Dr. Stelly was the Editor. However, that system created a problem for a volunteer Editor because he did not see manuscripts submitted to the journal, and only received copies of correspondence concerning the release of manuscripts to authors. Therefore, a new system was created in 1984, whereby manuscripts were submitted directly to the Editor and he assigned them to Technical Editors (ASA, 1985).

During this era of journal history, the Editorial Board was very concerned about the amount of time required for review and revision of manuscripts. In the 1984–1985 time period, the process took 7 to 9 mo for accepted manuscripts to move through the review process. An additional 4 to 5 mo were required before manuscripts were actually published, and thus, the total time from submission to publication averaged 11 to 14 mo. The Editorial Board focused their efforts on decreasing the amount of time between submission and the first communication back to the author because they believed that was the most crucial step. As a result, the time for this critical step averaged 9 to 10 wk, which was considered quite good, given that there was considerable mailing time involved. The Editorial Board also observed that delays in getting a manuscript to publication were frequently the result of authors not

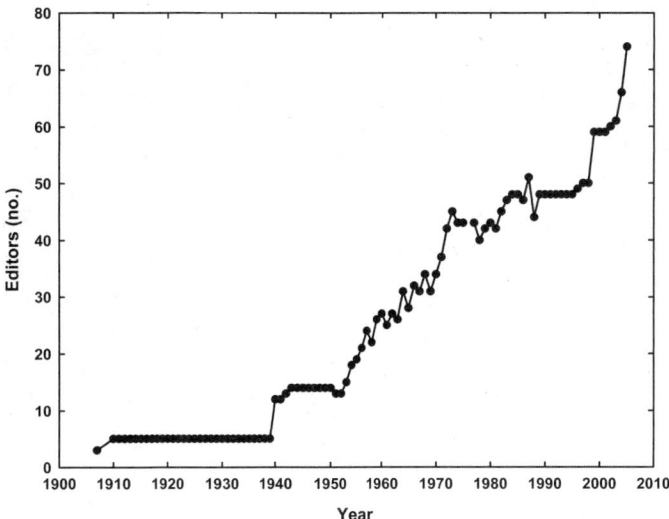

Fig. 2. The number of members on the Editorial Board of Agronomy Journal *each year from 1908 to 2006.*

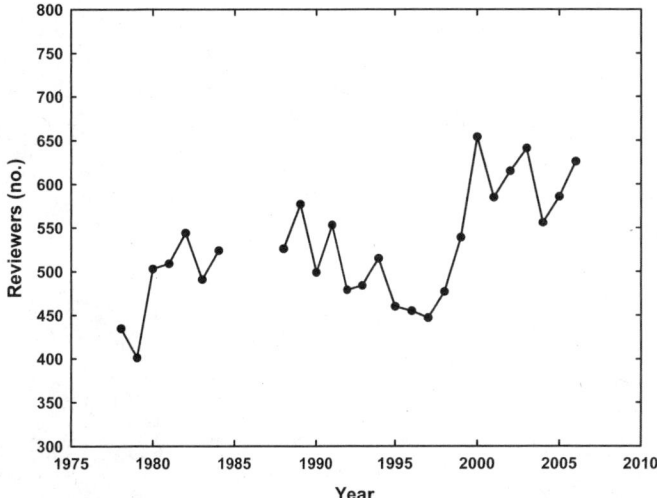

Fig. 3. Number of reviewers who peer reviewed manuscripts each year for the Agronomy Journal Editorial Board from 1978 through 1984 and from 1988 through 2006.

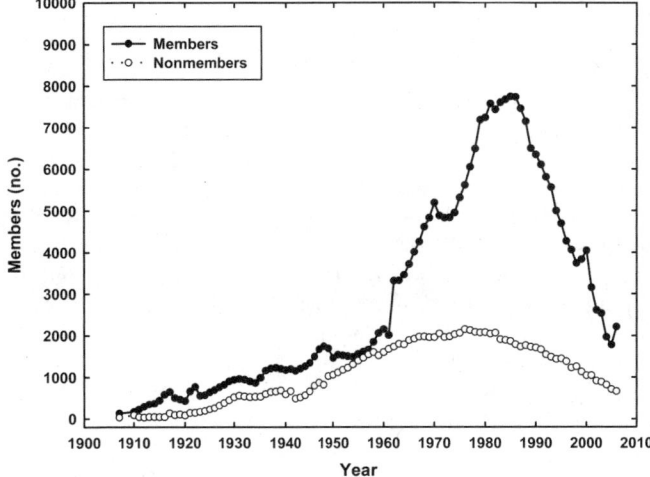

Fig. 4. The numbers of members and nonmembers who subscribed to Agronomy Journal from 1908 to 2006.

responding promptly after receiving their reviews. At the Editorial Board meeting held in 1988 in Anaheim, CA, the Editorial Board adopted the *3-mo rule* in which authors were allowed 90 d to complete a revision (ASA, 1989b). Shortening the time for authors to complete a revision from 6 to 3 mo was instituted to encourage authors to respond in a timely manner to editorial comments.

In response to a membership initiative, *Agronomy Journal* began to solicit review articles during this era. The Editor requested potential titles, subject matter areas, and authors from the membership, and 12 suggestions were received in the first round of solicitation. The invited review papers were subject to the standard review process. The goal was to publish two or three review papers during 1988 and at least two or three additional papers per year on into the future.

During this time period, *Agronomy Journal* developed a policy regarding the publication of models, and they were first published in a journal section called *Agroclimatology and Agronomic Modeling*. *Software Scene*, a place to publish new software, was also added and the first three of such papers appeared in the March-April issue of 1989.

After much discussion the Editorial Board agreed to divide the table of contents into categories to assist readers in finding papers specific to their interests. The initial five categories agreed on were: *Crops, Soils, Agroclimatology and Modeling, Notes*, and *Software Scene*. These categories first appeared in the January–February issue of 1990 (ASA, 1990). Later it was proposed that *Agroclimatology and Modeling* be split into two parts to help attract more modeling papers.

Interestingly, in fall 1987, the question was raised about the possibility of electronic manuscript submission. Guidelines already existed for submitting a manuscript on disk once it had been reviewed and approved for publication, but Managing Editor Bill Luellen reported that submission, review, and editing of papers online was still a few years down the road (ASA, 1988).

Fig. 5. Editors of Agronomy Journal *during its 100-year existence from 1908 to the present. From left to right, top to bottom. Carleton R. Ball, C.W. Warburton, Roscoe W. Thatcher, James D. Luckett, Maurice R. Haag, W. Charles Robocker, L.G. Monthey, Matthias Stelly, Dwayne R. Buxton, Gary A. Peterson, Jerry L. Hatfield, Kenneth A. Barbarick, Calvin H. Pearson, and Donald G. Bullock.*

1990–1995

Concern over the quality of the reviews and potential bias in the reviews because of name recognition of authors prompted a change in the review process. The journal reviewers, as a standard practice, were anonymous to authors, and comments were often received about the unprofessional nature of the comments. One of the roles of Associate Editors and Technical Editors was to remove any biased and unprofessional comments from reviewer comments. As an experiment, the review process changed to a completely double-blind review in which the author's names and institutions were removed from the manuscript (ASA, 1992). After 3 years, a survey was conducted about the review process and it was determined that the double-blind review process imparted increased confidence about the quality of the reviews such that this procedure became standard operating procedure.

During the 1990s there was ongoing discussion about the appropriate statistical tools to be used in agronomic research. This affected statistical analyses used in research papers and finally culminated in a more inclusive use of statistical approaches rather than a single method for mean separation. One of the major changes made was to modify the Instructions to Authors to accept all appropriate statistical methods for agronomic studies. This change increased the papers that addressed spatial aspects in field studies. One of the goals of this change was to foster more research papers describing innovative approaches for research and field-scale studies.

1996–2001

In 1996, 6 years before the advent of Manuscript Tracker,[1] we initiated electronic manuscript submission and review. Technical Editor Robert Lascano conducted the trial run and found that it took 1 to 5 d to identify reviewers and forward the electronic files, 55 d before the first review comments were electronically returned to the authors, and 87 d for the authors to electronically submit their revision. This initial electronic manuscript was accepted. We formally initiated electronic

---

[1] Manuscript Tracker *is an online, wed-based system for electronically submitting and reviewing manuscripts. Manuscript Tracker is used by authors, reviewers, and editors.*

Table 1. Editors, their affiliation, dates of service, and number of years of service during the 100 years Agronomy Journal *has been published.*

No.	Editor	Affiliation	Dates of service
1	Carleton R. Ball	Office of Cereal Crops and Diseases, Bureau of Plant Industry, USDA, Washington, DC (Ball et al., 1928)	1909–1914 (Throckmorton, 1941)
2	C. W. Warburton	Director of Extension Work, USDA, Washington, DC (Warburton, 1925)	1915–1921 (Throckmorton, 1941)
3	R. W. Thatcher	Director of Experiment Stations, Cornell University, Geneva, NY; Director of the New York Agricultural Experiment Station (Fuccillo, 1983; Thatcher, 1927)	1922–1927 (Throckmorton, 1941)
4	J. D. Luckett	Editor, New York State Agricultural Experiment Station, Geneva, NY (Luckett, 1927)	1928–1948 (Throckmorton, 1941; Fuccillo, 1983)
5	Maurice R. Haag	Assistant Extension Editor at Univ. of Wisconsin, Experiment Station Editor at Univ. of Wyoming, and Managing Editor of the Proceedings of the Soil Science Society of America (Fuccillo, 1983)	Jan. 1949–1952 (ASA, 1948a)
6	W. Charles Robocker	Assistant in Agronomy at the University of Wisconsin (ASA, 1952)	July 1952–April 1953
7	L. G. Monthey	Executive Secretary of ASA and Editor of What's New in Crops and Soils (ASA, 1953)	Apr 1953–1961
8	Matthias Stelly	Before becoming Editor he was at Louisiana State University, Agronomist at the Soil Testing Lab; was Editor and ASA Executive Secretary (Fuccillo, 1983)	Apr. 1961–1982 (Smith, 1980; Fuccillo, 1983)
9	Dwayne R. Buxton	Research Plant Physiologist, USDA-ARS, Ames, IA	1982–1983
10	Gary A. Peterson	Colorado State University, Dep. of Agronomy	1984–1989
11	Jerry L. Hatfield	USDA-ARS, National Soil Tilth Laboratory, Ames, IA	1990–1995
12	Kenneth A. Barbarick	Colorado State University, Dep. of Soil & Crop Sciences	1996–2001
13	Calvin H. Pearson	Colorado State University, Dep. of Soil & Crop Sciences, Agricultural Experiment Station	2002–2007

submissions in 1998 (ASA, 1998b); we received eight such submissions that year.

The membership requirement for publication was dropped in 1997. For 1998, we added the *Forum* section to allow authors to address a thought-provoking concept or idea to generate discussion among our readership. We also changed the *Notes* section to *Notes and Unique Phenomena* so that authors could present results for unusual occurrences such as crop response or other observations following hail damage (ASA, 1998b).

Several changes were implemented in 1999 through 2001. At the 1999 annual meetings in Salt Lake City, UT, the Editorial Board voted to start recognizing excellent reviewers (ASA, 2000; Barbarick, 2000). The award was titled the *Editors' Citation for Excellence in Manuscript Review* to recognize outstanding reviewers on an annual basis from various subject areas within the journal. A formal symposium-paper policy was approved at the 2000 Editorial Board meeting in Minneapolis, MN (ASA, 2001). The policy spelled out the procedure for publishing manuscripts in *Agronomy Journal* that were the result of symposia if they were found to have suitable subject matter (Barbarick, 2001). We debuted a new full-color cover, starting with the 1999 volume year (ASA, 1999). We began accepting production agriculture papers when the *Journal of Production Agriculture* was slated for termination in 2000 (ASA, 1999, 2000; Barbarick, 2000). With submission and eventual acceptance of production agriculture papers, we highlighted these articles in the table of contents from 2001 through 2006. In 2001, we published our first full color figure (Barbarick, 2001). Agronomy Journal was first posted online by Springer-Verlag with Volume 90 (1998) and then in 2000 with HighWire Press, Palo Alto, CA.[2]

2002–2007

On 13 Mar. 2002 we began using Manuscript Tracker for submitting, registering, reviewing, and tracking manuscripts submitted to *Agronomy Journal* (ASA, 2003). From this date forward all manuscripts submitted to the journal were logged into the Manuscript Tracker system. If manuscripts were submitted as paper copies, the Editor created a record in Manuscript Tracker and paper copies were handled as in previous times. Starting 1 Jan. 2004, the Editorial Board no longer accepted paper submissions (Pearson, 2004). Only electronic files of manuscripts were allowed after that date. At the time, the thought of eliminating the use of paper copies seemed a bit unrealistic; we were so used to dealing with paper. We have now been handling electronic files for several years and this has become very routine and normal. We questioned how readily some authors would adapt to using only electronic files, particularly scientists in developing countries who may not have adequate computer technology. This has not been a significant problem for most authors. They have adapted quite well to the Manuscript Tracker system.

Also during 2002, we initiated the preparation of news releases of selected papers from each issue of *Agronomy Journal* (Pearson, 2003). We contacted the authors of these selected articles and worked with them to prepare a news release of their paper. News releases were distributed through a distribution service of the American Association of the Advancement of Science to more than 4000 international science media. Articles from *Agronomy Journal* were also publicized on the web page of the Societies and in *CSA News* and the *News Flash* (semimonthly email sent out to members). Promoting the research published in *Agronomy Journal* continues to evolve and expand with the hope that the impact and contribution of the science found in *Agronomy Journal* will increase (ASA, 2005, 2006).

More formal associate editor appointments were instituted during 2002. This included a letter of appointment and certificate of appointment, both signed by the Editor. The certificates of appointment were suitable for framing and it was hoped that new Associate Editors would display them.

The use of electronic media continued to be more encompassing for publishing *Agronomy Journal*. On January 2003, PDF galley proofs (e-proofs) were sent to authors via the Internet (ASA, 2003). With the use of the Internet and Manuscript Tracker for submitting, registering, assigning, and tracking manuscript and the use of PDF galley proofs, a sizeable savings in postage costs has been realized.

Beginning in 2004, the traditional page charges were replaced with a publication fee. Some of the history of how page charges were assessed can be found in ASA (1979b, 1981, 1998a). The publication fee was on a per-manuscript basis, which was $450 for members and $700 for nonmembers (ASA, 2004a, 2004b; Pearson, 2005).

With the March–April 2004 issue of *Agronomy Journal*, authors began to select their own table of contents headings. As is the practice today, the authors are provided a list of headings and they choose the heading from the list that best describes the content of their manuscript. The procedure gives authors increased input into the publishing of their articles (Pearson, 2004).

After much work, largely through the efforts of Charles Roth from Purdue University, a seven-CD set of the back issues (Vol. 1–93,

---

[2] HighWire Press (http://highwire.stanford.edu/), a division of the Stanford University Libraries, hosts more than 1000 scientific journals. *Agronomy Journal* is available at http://agron.scijournals.org/.

1908–2001) of *Agronomy Journal* became available for purchase during 2004 (Pearson, 2005). Digitizing the journal was part of a larger effort to make CD sets for all Society journals. The seven-CD set for *Agronomy Journal* also included the *Journal of Production Agriculture* and the *Journal of Natural Resource and Life Sciences Education* (ASA, 2004b). Thus, all volumes of *Agronomy Journal* are now available in a digital medium.

Starting with the May-June 2005 issue, we began using a concept known as "continuous publication" (ASA, 2005, 2006; Pearson, 2005, 2006). Manuscripts are posted online each month with fully citable reference information. Continuous publication is an economical method for faster publishing of research papers, rather than waiting 2 mo before a paper issue of *Agronomy Journal* is received in the mail.

During the meetings in Salt Lake City in November 2005 the Editorial Board approved the addition of an Acquisitions Associate Editor (ASA, 2006; Pearson, 2006). Wesley Rosenthal from Texas A&M University at the Blackland Research Center was appointed on 22 Feb. 2006 as our first Acquisitions Associate Editor. The Acquisitions Associate Editor is responsible, in cooperation with the Editor and the Editorial Board, for soliciting content for *Agronomy Journal*, with a primary focus on obtaining review articles.

During the past 23 years, the length of articles published in the journal increased from an average of 4.6 pages in 1984 to 8.4 pages in 2006 (Fig. 6). Fuccillo (1983) noted that the average length of articles in Volume 50 (1957) was 3.7 pages. Thus, during the past 50 years (from 1957 to 2006) the average length of articles has more than doubled and over the past 23 years (from 1984 to 2006), the average article length has increased 83%. Pages charges for *Agronomy Journal* and most of the other society journals underwent considerable change in 1979 (ASA, 1979a, 1979b) and again in 1981 (ASA, 1981). In 1979, page charges were $30 per page for 4 journal pages or less. For articles over 4 pages, a $120 production charge was levied for each page and assessed in half page increments. In 1981, page charges were increased to $40 per page for 4 journal pages or less. For articles over 4 pages, the production charge was increased to $150 a page. It was part of the editorial culture to encourage papers to be short with the idea that they would be more readable. This was somewhat of a deterrent against long articles and encouraged authors to prepare concise papers during this time period. In 1998, a new page charge structure went into effect and there were no charges for the first 6 printed pages with a production charge of $165 per printed page beyond 6 pages of a manuscript (ASA, 1998a).

The overall reason for an increase in paper length is not certain, but we speculate that the additional length of many articles may be the result of scientists addressing more complex issues and the capability of collecting more data with today's instrumentation, thus more verbiage is needed to explain how the study was conceived, conducted, and to present and interpret the findings and impacts of those findings of these more complex and comprehensive research projects. We doubt that authors of today are any more or any less verbose than authors of earlier times. Another possible reason for longer papers is the publication fee. Authors have no financial deterrent when they write long papers. It is interesting to note that concern about the length of articles is not new. Editor M.R. Haag was concerned about a backlog of manuscripts that had developed during the early 1950s and he wrote in his annual report (ASA, 1951), "One of the steps toward dealing with the matter is insistence on shorter and more concise papers."

Data for international papers submitted and published in *Agronomy Journal* have only been collected in recent years (Table 2). International papers are valued in the journal and are needed for the contribution they make to the agronomic and natural resource sciences and to the viability of our publication. As an Editorial Board we continue to encourage the submission of high quality international papers and we seek to help international authors and others to prepare manuscripts that are suitable for publication in *Agronomy Journal*.

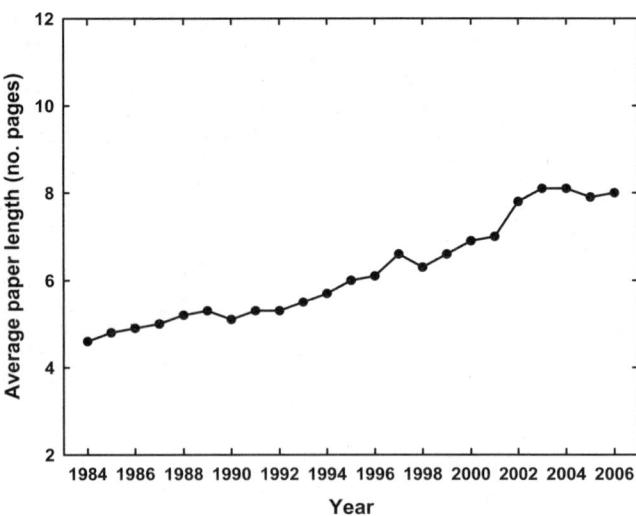

Fig. 6. The average length of articles published in Agronomy Journal *from 1984 through 2006.*

Some of the major changes that *Agronomy Journal* has experienced over the past 25 years are summarized in Table 3. As a current snapshot of *Agronomy Journal*, we published 60% of the manuscripts submitted in 2005. In both 2003 and 2004, we accepted 55% of the manuscripts submitted.

Recently, the use of metrics to compare peer journals has become more widely used. In a comparison of 48 peer journals in 2005, the impact factor of *Agronomy Journal* ranked 12th at 1.473 and the total citations for the journal ranked fourth at 6723 (Thompson Institute for Scientific Information, 2006).

WHERE DO WE GO FROM HERE?

Our biggest challenge may be to stay economically viable in the face of increasing pressure to allow full and free public access to articles. We also need to enhance *Agronomy Journal* as the preeminent source of agronomic information. Our journal is well-suited for the integrative, multidisciplinary research that granting agencies highly value today. We encourage and are attempting to work actively with our international colleagues and others to help them prepare high-quality submissions of their research results. Too often, good science is not recognized because the writing is substandard.

Of continuing importance will be for us to be faster at completing reviews, revisions, and publishing articles in *Agronomy Journal*. We

Table 2. International papers submitted and international papers published in Agronomy Journal *in recent years.*

Volume year	International papers submitted†	International papers published‡
	————————%————————	
2000	–	31
2001	–	34
2002	36	23
2003	30	24
2004	28	22
2005	38	23
2006	39	26

† Percentage of the total number of manuscripts submitted that was international.
‡ Percentage of the total number of manuscripts published that was international.

Table 3. Significant changes affecting Agronomy Journal occurring from 1983 through 2007.

Year	Volume	Significant change by year
1983	75	75th Anniversary Year (ASA, 1983b)
1984	76	Changed heading from "Literature Cited" to "References" (ASA, 1984b)
1985	77	Changes begin for manuscripts to be submitted directly to Editor and not Headquarters (ASA, 1985)
1986	78	First personal computer purchased for use by the Headquarters Editorial Staff (ASA, 1986)
1987	79	First two software papers submitted to the journal (ASA, 1987)
1988	80	The possibility of submitting manuscripts electronically discussed (ASA, 1988)
1989	81	Time allowed for author revisions of papers shortened from 6 to 3 mo (ASA, 1989b). Beginning in 1989, other society journals can be substituted for Agronomy Journal to satisfy the ASA membership journal requirement (ASA, 1989a)
1990	82	New divisions for the table of contents debuts with the first issue of 1990 (ASA, 1990)
1991	83	The words "An International Journal" added to the cover starting with the Jan-Feb 1991 issue (ASA, 1991)
1992	84	Double-blind reviews approved for a 2-year trial (ASA, 1992)
1993	85	Discussed perception that the journal is too conservative and thus loses valuable submissions (ASA, 1993)
1994	86	Sheridan Press begins printing Agronomy Journal and the other five Society journals (ASA, 1994)
1995	87	All journal manuscripts are now edited using a computer (ASA, 1995)
1996	88	Initiated electronic manuscript submissions
1997	89	Dropped membership requirement for publication (ASA, 1998b)
1998	90	Forum section initiated; initiated "fast-track" manuscript reviews; changed the "Notes section to "Notes and Unique Phenomena" (ASA, 1998b); Online with Springer-Verlag
1999	91	New Agronomy Journal color cover debuts along with new headings in the Table of Contents (ASA, 1999); The words "… of Agriculture and Natural Resource Sciences" added to the cover starting with the Jan-Feb 1999 issue (ASA, 1999). Began accepting "Production Agriculture" manuscripts coinciding with the termination of the Journal of Production Agriculture
2000	92	Began highlighting "Production Papers" in the table of contents; first color figure published
2001	93	Online with HighWire Press, Palo Alto, CA
2002	94	Began using Manuscript Tracker for online manuscript submissions (ASA, 2003)
2003	95	An author/subject index is no longer published [Agron. J. 95(6):iii]
2004	96	Paper manuscripts are no longer accepted; began using publication fees instead of page charges; largest volume ever published (1828 pages); CD back-issue set available (Pearson, 2004; ASA, 2003)
2005	97	Continuous publication begins with the May-June 2005 issue (ASA, 2005)
2006	98	An Acquisitions Associate Editor is added to the Editorial Board; May-June 2006 is the largest issue ever published at a hefty 456 pages (ASA, 2005)
2007	99	Publication fees and page charges undergo review; "Production Papers" highlighting discontinued

also need to think globally and inclusive in all related aspects of our scientific publishing.

Agronomy Journal also is important as a record of the business, functions, and accomplishments of the American Society of Agronomy and its members. This will continue to be important in the future as it has been in the past.

We would be well served as an Editorial Board to function in a more customer-oriented approach. This does not mean we put less emphasis on our scientific values, but in performing our editorial duties we are, in fact, performing service to our customers (the authors, subscribers, and readers) and we should perform all those duties and services in the best possible manner.

The journal will not survive and thrive without qualified reviewers and editors. People should seek to acquire skill sets for reviewing and editing and then volunteer to be reviewers and editors. As Laude et al. (1962) and Gary A. Peterson (ASA, 1997) have both noted that our journals are key to our societies. Thus, our publications are critical to the continuation of our societies.

Agronomy Journal is now 100 years old. Its success, its longevity, its scientific value—can be attributed to the dedication of many people who have contributed as authors, reviewers, editors, and others. We are honored to have played a small part in the 100-year history of *Agronomy Journal*. Other authors, reviewers, and editors will follow us. We fully expect the future of *Agronomy Journal* will be even more exciting, rewarding, challenging, and valued as the past 100 years. We eagerly look forward to the next 100 years of *Agronomy Journal*.

REFERENCES

ASA. 1948a. Maurice R. Haag named Agronomy Journal Editor. J. Am. Soc. Agron. 40:1142–1143.

ASA. 1948b. Agronomic affairs. J. Am. Soc. Agron. 40:1148–1154.

ASA. 1951. The forty-third annual meeting of the American Society of Agronomy. Agron. J. 43:626–628.

ASA. 1952. Agronomic affairs. Agron. J. 44:390–393.

ASA. 1953. Agronomic affairs. Agron. J. 45:221–224.

ASA. 1979a. Executive Vice-President's report. Agron. J. 71:188–192.

ASA. 1979b. Reports of divisions, branches, and committees. Agron. J. 71:193–214.

ASA. 1981. ASA Executive Committee meetings. Agron. J. 73:202–214.

ASA. 1982. Reports of divisions, branches, and committees. Agron. J. 74:238–251.

ASA. 1983a. Reports of divisions, branches, and committees. Agron. J. 75:375–391.

ASA. 1983b. Diamond jubilee of the American Society of Agronomy. Agron. J. 75:397–400.

ASA. 1984a. ASA Executive committee meetings. Agron. J. 76:324–331.

ASA. 1984b. Reports of divisions, branches, and committees. Agron. J. 76:335–346.

ASA. 1985. Reports of divisions, branches, and committees. Agron. J. 77:334–346.

ASA. 1986. Reports of divisions, branches, and committees. Agron. J. 78:391–401.

ASA. 1987. Reports of divisions, branches, and committees. Agron. J. 79:399–410.

ASA. 1988. Reports of divisions, branches, and committees. Agron. J. 80:361–373.

ASA. 1989a. ASA Headquarters report, 1988. Agron. J. 81:322–324.

ASA. 1989b. Reports of divisions, branches, and committees. Agron. J. 81:335–344.

ASA. 1990. Reports of divisions, branches, and committees. Agron. J. 82:376–384.

ASA. 1991. Reports of divisions, branches, and committees. Agron. J. 83:443–453.

ASA. 1992. Reports of divisions, branches and committees. Agron. J. 84:281–291.

ASA. 1993. Reports of divisions, branches and committees. Agron. J. 85:443–453.

ASA. 1994. Reports of divisions and committees, 1993. Agron. J. 86:377–392.

ASA. 1995. Reports of ASA divisions, branches, and committees. Agron. J. 87:294–308.

ASA. 1997. Reports of ASA divisions, branches, and committees. Agron. J. 89:302–316.

ASA. 1998a. ASA Headquarters report, 1997. Agron. J. 90:242–255.

ASA. 1998b. Reports of ASA divisions, branches, and committees. Agron. J. 90:256–270.

ASA. 1999. Reports of ASA divisions, branches, and committees 1998. Agron. J. 91:504–521.

ASA. 2000. Reports of ASA divisions, branches, and committees, 1999. Agron. J. 92:560–575.

ASA. 2001. Reports of ASA divisions, branches, and committees, 2000. Agron. J. 93:687–707.

ASA. 2003. Reports of ASA divisions, branches, and committees, 2002. Agron. J. 95:734–761.

ASA. 2004a. ASA Headquarters report, 2003. Agron. J. 96:857–872.

ASA. 2004b. Reports of ASA divisions, branches, and committees, 2003. Agron. J. 96:873–893.

ASA. 2005. Reports of ASA divisions, branches, and committees, 2004. Agron. J. 97:998–1011.

ASA. 2006. Reports of ASA divisions, branches, and committees, 2005. Agron. J. 98:840–848.

Ball, C.R., H.L. Shantz, and C.F. Shaw. 1928. Median terms in adjectives of comparison. J. Am. Soc. Agron. 20:182–191.

Barbarick, K.A. 2000. Letter from the Editor. Agron. J. 92:189.

Barbarick, K.A. 2001. Letter from the Editor. Agron. J. 93:261.

Fuccillo, D.A. 1983. The 75th publication year of Agronomy Journal. Agron. J. 75:413–417.

Laude, H.H., M.F. Miller, J.D. Luckett, G.G. Pohlman, D.S. Metcalfe, W.H. Pierre, and E. Truog. 1962. History of American Society of Agronomy: First fifty years—1907 to 1957. Agron. J. 54:57–69.

Luckett, J.D. 1927. Methods of graphic representation of experimental data. J. Am. Soc. Agron. 19:27–40.

Lyon, T.L. 1933. History of the organization of the American Society of Agronomy. Agron. J. 25:1–9.

Pearson, C.H. 2003. Letter from the Editor. Agron. J. 95:231–232.

Pearson, C.H. 2004. Letter from the Editor. Agron. J. 96:319–320.

Pearson, C.H. 2005. Letter from the Editor. Agron. J. 97:343–344.

Pearson, C.H. 2006. Letter from the Editor. Agron. J. 98:229–230.

Smith, D.C. 1980. Development of the American Society of Agronomy, 1958–1977. Agron. J. 72:227–240.

Thatcher, R.W. 1927. Should the results of agronomic research be published in bulletins, or in scientific journals, or both? J. Am. Soc. Agron. 19:2–7.

Thompson Institute for Scientific Information. 2006. ISI web of knowledge. v. 3.0. Available at http://portal.isiknowledge.com/portal.cgi?DestApp=JCR&Func=Frame [verified 25 May 2007]. Thompson Institute for Scientific Information, Stamfort, CT.

Throckmorton, R.I. 1941. History of the American Society of Agronomy. J. Am. Soc. Agron. 33:1135–1140.

Warburton, C.W. 1925. Taking agronomic research to the farmer. J. Am. Soc. Agron. 17:757–764.

Resident Education

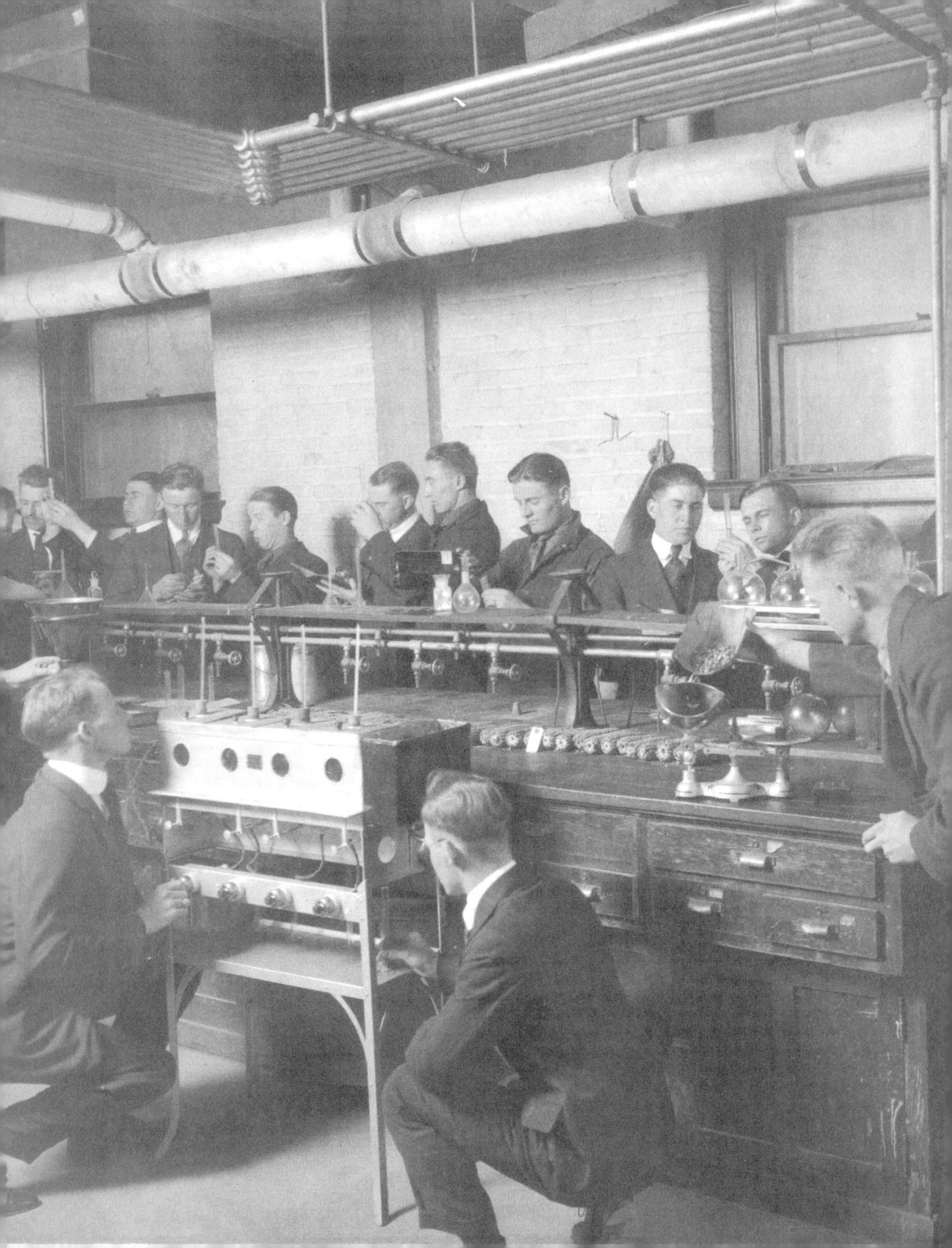

Development and Evolution of Resident Education in the American Society of Agronomy

At the 100th birthday of the American Society of Agronomy (ASA) we celebrate the development and evolution of the strong ASA educational programs for professionals and students of the agronomic sciences. In preparing this chapter we relied heavily on minutes from Board meetings, annual reports of educational committees, some perusal of the ASA journals, program booklets for annual meetings, input from individual educators, and our personal experiences as participants. We hope that readers will be as impressed as we have been with the commitment of ASA to serving the needs of educators and students by supporting programs in the areas of service, outreach, and sharing of research and information

M. H. Milford
Department of Soil and Crop Sciences
Texas A&M University, College Station

D. T. Smith
Department of Soil and Crop Sciences
Texas A&M University, College Station

EARLY FOUNDATIONS IN AGRONOMIC EDUCATION

In his presidential address to ASA, W.M. Jardine (1917) stressed the importance of agronomic education for a sustainable agriculture. He noted that World War I was creating a global crisis for food and fiber supply due to shortfalls in domestic production and the inaccessibility of critical products. Jardine (1917) was particularly concerned about the supply of young agronomists for the future. He emphasized the need for more "college men" trained in crop selection (referring to crop breeding) and management of soils, as well as the agronomic courses that built upon recent advances in chemistry, physics, and other basic sciences. Jardine (1917) advocated that the agricultural challenges in the future could be solved by trained men with B.S. and advanced degrees. Although numerous professors and colleges were already well-engaged in agronomic education, this was one of the earliest published records highlighting agronomic education as a strategic issue for the profession. It is noteworthy that this Kansas agronomist continued to provide vision, later becoming U.S. Secretary of Agriculture. It is reasonable to speculate that major advances in U.S. agriculture in the following decades had their origins in the work of Dr. Jardine.

Early Educational Emphasis

Few articles on educational issues were published in the early volumes of the *Journal of the American Society of Agronomy* (JASA). Key issues and academic programs were recorded primarily through annual reports from committees. In the formative years of ASA, professors commonly discussed the academic content of agronomy courses. Concerns about the adequacy of courses resulted in several conferences and workshops to establish academic expectations for undergraduate students. In the early 1920s, attention focused on standardizing curricula and course content, particularly in introductory courses. Committees were formed to address teaching of field crops and soils, as well as the possible involvement of students in ASA. Agronomic educators conducted surveys to obtain information about course content. For example, a 1926 committee on crops gathered information on introductory courses, predominately for first and second year students. The survey showed that 67% of the institutions had adopted a standardized course outline, usually with modifications to fit local or regional needs.

Corresponding author
M.H. Milford, Dep. of Soil and Crop Sci., Texas A&M Univ. College Station TX 77483-2747 (email: MMilford@suddenlink.net).

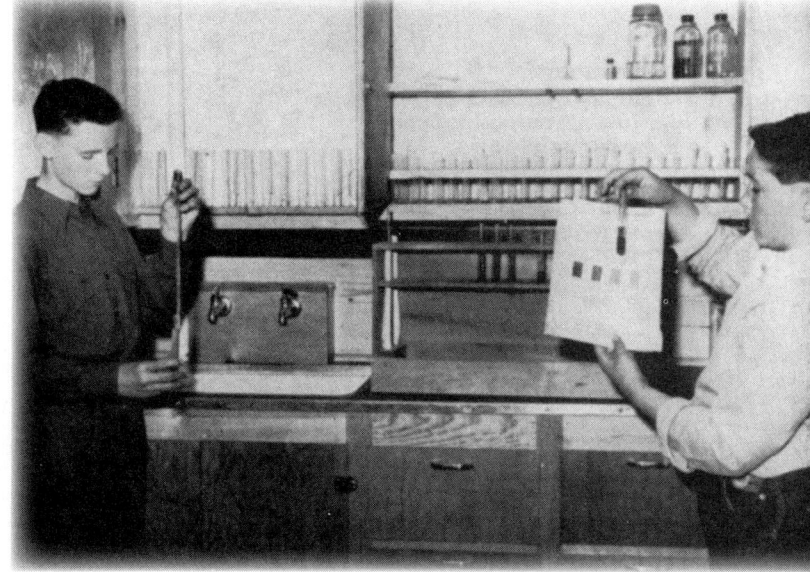

A photograph featured in a 1950 Crops and Soils *article on attracting young scientists to the field of soil science.*

Left: An agronomy class in the early 1900s. Photo courtesy of the University of Wisconsin-Madison Agronomy Department.

Section page: Students engrossed in a a soil judging contest, a popular student activity.

A few respondents were critical of such an approach, indicating that the course outline was either too general or did not fit their needs. Some were not interested in modifying their course to fit local needs. Similarly, a questionnaire was sent to all U.S. and Canadian colleges regarding soils courses and teaching methods. Round table forums at meetings featured debates on whether agronomic instruction was a science or simply an application of botany and other sciences and outlined the essential skills and proficiencies that should be imparted to students, a forerunner of today's "learning outcomes."

Before 1920, the few papers published in JASA focused on descriptive botany, crop responses, and nomenclature. Up to the Great Depression in the 1930s, published papers focused on ways to achieve quality, consistency, and academic adequacy among colleges throughout the United States. For example, soils courses at Iowa State College were summarized (Brown, 1916) in an effort to establish a benchmark and level of expectation. In 1920, 16 land grant colleges participated in a conference at Lexington, KY on the teaching of soils (Karraker, 1920). Conference topics included the presentation of scientific principles of soils via classroom and laboratory experiences. Subsequently, six papers about teaching methods, content, and academic programs were published in 1921 [JASA 13:49–81]. While the first edition of *The Nature and Properties of Soils* (Lyon and Buckman, 1922) provided an excellent benchmark text on soils, reviewers noted that the subsequent edition in 1930 was a great advancement in teaching soil science principles to undergraduate students.

Organizational Structure

The evolution of ASA divisions was summarized by Pohlman (1962). The first program committee for an annual meeting of ASA (1908) consisted of a soils representative and a crops representative (Table 1), but there was no separation within the meeting program until 1924. Committees on soils teaching and crops teaching existed during this time and have continued through the years. The first organizational units were a soils section and a crops section that were recommended in 1930 and made a part of the constitution in 1932 (Pohlman, 1962). A standing committee on agronomic education was established by 1930 and continued for more than a decade. In 1948, an Agronomic Education Division was formed. Subsequently, during the next three years, the Resident Teaching, Extension Participation, and Student Activities Sections were created. With the passage of revised ASA by-laws in 1952, major structural changes occurred. Six divisions of the Soil Science of America (SSSA), six crop science divisions, and a 13th division in agronomic education were codified. The Agronomic Education Division was related to both groups but operated independently. The list of 1953 officers in *Agronomy Journal* included those for Division XIII as well as for three subdivisions: XIII-A, Resident Teaching; XIII-B, Extension Teaching; and XIII-C, Student Activities. In 1963, the Division became A-1, and its associated subdivisions were A-1a, A-1b, and A-1c. In 1964, Resident Teaching, subsequently called Resident Education, received division status and was given the A-1 designation. Student Activities became a subdivision with an A-1a designation. Although both A-1 and A-1a were structurally associated with ASA, both had relations with Crop Science Society of America (CSSA) and SSSA. In 2004, in a move to gain greater prominence for student programs, A-1a was dropped and transformed into a new organization, Students of Agronomy, Soils, and Environmental Sciences (SASES), that was associated with all three Societies.

Educational Trends

In 1935, the standing committee on Education in Agronomy approached the American Council on Education to include applied agricultural fields in the listing of studies in graduate education. The goal was to call attention to the need for graduate instruction in foundation sciences, fields of agronomy, and the need to recruit superior students for graduate study.

After the Great Depression and during WW II, ASA established a committee in 1943 to consider war and post-war adjustments to academic programs. One of the outgrowths focused on establishing more suitable courses for servicemen once they returned from the armed forces. Federal support and the return of GIs to the classroom had a significant impact on higher education in the agronomic sciences. Many of the returnees completed a B.S. degree, earned a Ph.D. in agronomy, and subsequently entered the professional work force with a high degree of maturity. This cadre of professional agronomists advanced into positions of leadership in both the public and private sectors, establishing a stronger scientific basis and transfer of technology in U.S. agriculture. In the late 1970s and early 1980s, however, just as enrollment of students majoring in agronomy peaked, agronomists of this era retired from the work force, leaving substantial gaps in the scientific staffing of many agronomic programs.

Within the Resident Teaching Section in 1958, a committee on training of agronomists had subcommittees on curriculum development, recruitment, faculty and facilities, employment, and improvement of teaching. Symposia, papers, and other actions in succeeding years reflected studies and recommendations coming from the activities of these action committees and their successors.

Activities within the Resident Teaching Section of the Agronomic Education Division accelerated from the late 1950s with the activity of previously mentioned subcommittees and their successors. After the Resident Teaching Division A-1 was established in 1964, numerous advances occurred. Subcommittees and results of their work included:

Table 1. History of educational structure and activities in ASA.

Year	Event
1907	Program committee for 1908 meeting consisted of one crops and one soils member
1920s	Committees established: crops teaching methods, soils teaching methods
1921	Committee on national organization of college students appointed
1923	Student organization recommended, but left to student initiative
1930	Standing committee on agronomic education established
1932	Committee on organization of student sections appointed
1935	First meeting of student organization held with that of ASA in Chicago
1948	Agronomic Education Division established Section I Resident Teaching Section II Extension Participation Section III Student Activities
1950	Students start holding national meetings at ASA meeting location
1952	Agronomic Education Division (XIII) given a roman numeral designation Section XIII-A Resident Teaching Section XIII-B Extension Teaching Section XIII-C Student Activities
1964	Resident Teaching gained Division status with code designation (A-1) A-1a Student Activities became a Subdivision Extension Teaching Division (A-4) formed
1965	Resident Education Division (A-1) renamed A-1a Student Activities Subdivision Extension Education Division (A-4)
1972	*Journal of Agronomic Education* established
2004	Students of Agronomy, Soils and Environmental Sciences (SASES) formed with ASA, CSSA, and SSSA

by-laws, teaching improvement (establishing the *Journal of Agronomic Education*; JAE), educational resources (creating banks of test questions), international education in agronomy (developing a symposium and papers), reference materials for teachers (creating a film inventory), manpower for agronomy, educational exhibits, national senior award (developing student recognition programs), national photography contest, and career programs. A final recommendation from the 1969 meeting was to reduce the number of committees since many had achieved their major goals.

During the past quarter century the Resident Education Division hosted many symposia on educational issues at the annual meetings. For example, between 2001 and 2006, there were an average of 35 oral and poster presentations as a result of symposia and special themes. Featured topics included outcome assessments, curricula, interactive learning, instructional innovations and teaching resources.

Regional Groups and Student Involvement

By 1912 ASA had recognized the need for regional agronomic organizations (Pohlman, 1962). The original ASA constitution was revised to provide geographical sections. Eventually four regional branches persisted: North Central (formerly Corn Belt), Northeastern, Southern, and Western. Some of the Regional/Branch meetings included professional activities for students, similar to the contests and events at national meetings. Participation in some regional contests, such as crops judging and soil judging, has been a prerequisite for competing at the national level. Regional student meetings, regularly held in the North Central Region, beginning in 1973, and more recently in the Southern Region, include professional activities focused on agricultural and natural resources, as well as social events. Today, many students are particularly active in regional meetings because of the close proximity to their home campus.

Gender

While perhaps inaccurate, a survey report in 1930 by the Education in Agronomy Committee indicated that all holders of B.S., M.S., and Ph.D. degrees in agronomy were men. The first woman joined ASA in 1937 (M. McIntosh, personal communication, 2007). Collins and Pesek (1983) summarized the status of women in science in the USA. In 1920, 15% of all Ph.D. degrees were granted to women, which grew nominally to 30% by 1980. Growth was attributed to several factors: societal changes after WW II when more women entered the work force, the civil and equal rights movements of the 1960s, and equity hiring practices by public agencies and universities. Collins and Pesek (1983) noted that the number of women was still critically low in many fields, including agricultural sciences, particularly since few role models existed for young women until the 1980s. For example, in 1980, women made up less than 8% of the teaching faculty in soil microbiology. The first meeting of the Women in Agronomy committee occurred in 1981, with an estimated attendance of 50. In 1986, there was a symposium on "evaluation and promotion of the young professional" sponsored by Resident Education (Division A-1) and Women in Agronomy, which highlighted the needs and issues for young professionals. By 2006 participation of women grew to 130. In 2005, the total resident ASA membership of women consisted of 10.3, 34.7, and 32.4% as active, graduate student, and undergraduate student members, respectively (data from ASA Headquarters staff). These data clearly indicate the increased involvement of women as members and leaders within ASA. The increased level of participation by women was evident when the 1998 soil judging contest was won by an all-female team from Texas A&M University, which was an all-male institution until 1963. At the last two national meetings, women represented nearly one-half of the participants in job interview and graduate school/career enhancement workshops provided by the Societies. On

Victorious regional soils contestants, 1986.

the basis of preregistration numbers, 40% of the students attending the meetings in 2006 were women.

Minority Recruitment

In 1997, the ASA Executive Committee established an Agronomy Club Mentoring Program and allocated $10,000 to increase the number of affiliated student chapters and to encourage exchange between proposed, new, or revitalized clubs and established student chapters. The program encouraged intercampus exchanges between 1862 Land Grant Institutions, 1890 Institutions, 1994 Tribal Colleges, and other institutions. From 1998 to 2001, seven proposals were funded. In 2006, CSSA established a Golden Opportunity Institute program to attract and retain talented students in crop science. Travel support, special programs, and mentoring were provided for 15 recipients to attend the annual meeting.

K–12 and Outreach

The Resident Education Division has hosted many symposia on educational issues during the annual meetings over the past quarter century. ASA made special overtures to reach kindergarten through 12th grade (K–12) teachers in the vicinity of annual meetings. The goal was to provide more avenues for exposing youth to the challenges and opportunities in agronomic sciences. For example, the 1998 annual meeting involved regional science teachers in the Baltimore area to expose them to lessons and laboratory exercises on agronomic sciences to help attract high school students to careers in agronomy. These local outreach sessions were continued at the next two annual meetings, but were discontinued because feedback was not particularly strong and the impact seemed minimal.

PUBLICATIONS

Publication of educational materials has been a challenge for members since the inception of ASA. Six education-oriented papers appeared in 1921 (JASA 13:49–81), but this was rarely eclipsed in the next two decades. Between 1949 and 1953, six to eight education papers were presented annually at ASA meetings. From 1963 through 1971, Agronomic Education Division/Section headings appeared in the table of contents of *Agronomy Journal* (AJ), but the number of education papers published in each volume dwindled from 12 in 1963 to only two per year in 1966 and 1967. Even fewer were published from 1968 through 1971.

In 1969, the Teaching Improvement Committee of Division A-1 investigated several possibilities to enhance publication of educational papers. The committee considered four options: seeking more freedom for AJ editors to publish educational papers, devoting a special issue of the AJ to teaching, establishing publication relationships with other journals, and establishing a new journal within ASA. The last option was pursued with success, with the establishment of the *Journal of Agronomic Education* (JAE).

Journal of Agronomic Education

With the launch of JAE in 1972, a new avenue for publication became available. While this new journal was welcomed by educators, there had been some prior advantage in having research and education articles published and disseminated together in the same journal. Nonetheless, this new outlet catered to educators and increased the number of oral presentations at meetings and, subsequently, the number of published papers.

The content of educational papers has evolved during the past six decades. Clearly, by the 1970s, papers had shifted from the early concerns about academic content of college courses in the pre-war years to strategies for reaching students who had been raised on pre-school educational television. For example, the first five volumes of JAE included papers on audio-tutorial instruction, demonstrations, electronic scoring of exams, environmental content, student evaluations, and graduate education. In the early 1980s, attention shifted to testing, field trips, use of videotapes, programming, and experiential learning. Problem solving and writing skills were added to the mix in the early 1990s. Recurrent themes over the years included laboratory exercises, testing, surveys, alumni occupations and reflections on their education, and student backgrounds, especially regarding rural and urban experiences.

By 1992, since the JAE continued to incur a deficit after maximizing circulation among ASA members, the journal's scope was broadened, and it was renamed the *Journal of Natural Resources and Life Sciences Education* (JNRLSE) (Graveel, 1992). Committees worked with the editor to broaden circulation and increase outreach to K–12 teachers. Papers dealing with environmental sciences, distance education, ethics in science and teaching, and K–12 science instruction became common place (Ernst and Graveel, 1996). By 2000, each issue contained at least two articles on demonstrations, lab exercises, or attendance motivations. "Druger's Notebook" provided by Dr. Marvin Druger, Past President of the Natural Science Teachers Association, became a regular front piece, with articles on techniques for successful teaching of science. In more recent times, presentations at ASA meetings and subsequent papers in JNRLSE have included teaching methodologies, strategies for involving "learners" in the educational discovery process, web-based courses, case studies, student teams, distant education, and career opportunities for agronomists. Today, JNRLSE is considered as a key educational journal for K–16 educators.

A review of other agronomic journals, published between 1930 and 1985, indicated that most of the educational literature was published in JASA, AJ, or JAE. The NACTA Journal, published by the North American Colleges and Teachers of Agriculture organization, attracts some agronomic papers. Over the decades, the *Soil Science Society of America Journal* (SSSAJ) published papers on soil judging, soils training/certification for students, and changes in judging contests. The *Journal of Production Agriculture* (JPA), published between 1988 and 1999, regularly contained papers on adult education, extension initiatives, technology transfer, and adaptive research. However, JNRLSE is now well established as the primary repository for papers on undergraduate education and teaching. In 1998, it began publishing papers electronically throughout the year, with a print compilation of all articles prepared at the end of each year.

Resident Education and Educators

The ASA provides a strong focal point for professional educators. In discussing this topic with several past chairs of A-1, colleagues cited

Ensuring a steady stream of future agronomists has been an ongoing theme in ASA's educational efforts.

multiple facets of the A-1 division, including presenting and hearing of papers, participating in symposia, conferring with colleagues on contemporary issues, visiting exhibits and posters, participating in committee meetings, and interviewing candidates for prospective jobs. James McKenna, Virginia Tech University, reflecting on 30 yr of participation in A-1 sessions, identified these benefits: obtaining a spirit of renewal from peers teaching at other universities, exchanging ideas on student counseling, regaining a mutual enthusiasm about teaching, returning to campus with new ideas, developing a strong zeal to do better, and feeling more creative. He also listed advancements in teaching methods highlighted during A-1 meetings: the advent of computers, Power-Point presentations, use of animation in lectures, case studies, on-line education, embracing environmental issues, use of global agronomic issues in the classroom, incorporating scientific advances into curriculum, and international exchanges. Vern Cardwell, University of Minnesota, noted the efforts to reach downstream into K–12 education and engage secondary teachers in our professional society and publications, and to sustain JNRLSE in the face of financial shortfalls and declining support for the academic mission at Land Grant institutions. The ASA members have sustained the A-1 division, and in doing so, the Society has maintained a position of leadership in teaching and collegiate education, while educational divisions in some other organizations have been redefined, dissolved, or merged over the years.

UNDERGRADUATE STUDENT PROGRAMS

Early History

Starting in the 1920s, a succession of committees and events eventually led to the present student group within the Society (Metcalfe, 1958, 1962). The development of a student organization arose from crop judging contests being held annually in Chicago and Kansas City. Although the interest in forming a national student organization was gaining favor within ASA, the primary impetus resided with those faculty advisors and students who were participating in annual crops judging contests. A 1923 ASA committee on national organization of college students recommended the formation of a student group, but the effort languished for a decade since there was no formal action, and ASA leadership felt the initiative should be student-based (Table 1) The first initiative came from some faculty advisors and students at the annual crops judging contests in Chicago where a few unidentified faculty members associated with the judging contests quietly guided student efforts to launch the formal involvement of students with ASA.

In 1932, a student committee met during the Corn Belt Section meeting of ASA, and in 1934 student representatives from Nebraska, Kansas, Illinois, Minnesota, Iowa, and Oklahoma (the first states to form student clubs) met at the Chicago crops judging contest to plan a one-half day session for the 1935 ASA meeting in Chicago. ASA reports document that the first annual meeting of a national student organization was held in Chicago on December 1, 1935 and was attended by 27 students representing five institutions. At this meeting, students decided to continue to meet annually, generally in conjunction with the International Crops Judging Contest in Chicago.

By 1936, 13 college student groups and teams (Table 2) were participating in meetings and contests annually in Kansas City and Chicago. By the 1941–1942 academic year, 23 student sections (at that time local student groups were called sections instead of clubs) had been recognized, and ASA distributed certificates of membership to 500 students. From 1943 to1947 the student section was inactive due to World War II. Participation of faculty was also curtailed due to gasoline rationing and travel restrictions.

Post World War II

Following the war, ASA revitalized student activities and established a standing committee on student sections. By 1947, 17 of 23 former student chapters were reactivated, with nearly 600 members, 20% more than before the war. Senior students who would not be eligible to hold office in the next year were appointed as temporary officers and asked to call a meeting of representatives in Chicago on November 29, 1947 to elect officers for the coming year. Representatives from nine student sections elected officers for the national student section, established a newsletter, assessed membership fees of 15 cents per semester, and set up an educational program for the next meeting. Their goal was to establish an agronomic student group or agronomy club at each U.S. institution that offered a baccalaureate degree in agronomic sciences.

By 1948 there were 22 student clubs, with a national membership of 720, which grew to 30 clubs with 1287 members by 1950, when the student section met for the first time with ASA at a location other than Chicago. Student Activities became a section in 1949 under the newly formed Agronomic Education Division, which then involved more than 1000 members in 25 campus clubs.

The Agronomic Education Division consisted of three sections: Resident Teaching (I), Extension Education (II), and Student Activities (III). In 1964, Resident Teaching (subsequently called Resident Education, A-1) and Extension Education (A-4) were established as separate divisions, with Student Activities as a subdivision (A-1a) of A-1. This organizational arrangement remained essentially unchanged for four decades, with some provisions for committees to oversee student activities. While Resident Education is a division within ASA, an ASA–CSSA–SSSA committee, Coordination of Resident Education Activities (ACS-524) provides oversight and guidance on matters of common interest.

Change to a Contemporary Name

Of recent but historical importance was the movement of the student organization from a Subdivision (A-1a Student Activities) within ASA to a direct affiliated relationship with ASA–CSSA–SSSA. Advisors from each of the three Societies now serve as the oversight committee for the students. The name of the student group was changed to Students of Agronomy, Soils, and Environmental Sciences (SASES). The major purpose was to broaden the appeal to student interests now encompassed in many related programs at land grant universities. This name change, in response to more diverse interests of students and the growing non-agricultural market for agronomic sciences, also met the objective to increase ASA–CSSA–SSSA involvement with students, broaden student base and club interest in affiliation with the Societies, and to potentially increase Society membership as students graduated. Many student members reported that peers on their campuses were definitely interested in contemporary topics, such as environmental stewardship and land management. By 2006, there were early indications of increased student membership and club affiliations with ASA, although the number of active clubs was no greater than it was in 1977 at the peak of undergraduate enrollment in agronomic programs. In 2006, the names of campus clubs were reviewed and predominantly included one or more of the following: "agronomy" (65%), "soils" (23%), "crops" (10%), "plants" (10%), and others such as "agricultural", "environmental", "field and furrow", "range", or "ranch".

From the beginning, student officers planned programs and activities for national meetings, with the help of faculty advisors. Today, officers keep campus clubs and members informed by newsletters and a website that provides news of upcoming national and regional agronomy meetings, contest information and rules, and information for potential new student clubs and members. Undergraduate students are active members of ASA, CSSA, or SSSA, in addition to being members of their campus clubs.

Student Programs at Annual Meetings of ASA

The 1935 student program, held with ASA in Chicago, was one-half day of discussions about club activities and judging contests. In other years, before 1950, national meetings were held in conjunction with the crops judging contests rather than with ASA. Student programs have grown into three days of concentrated activities with a focus on professional development. In the past decade student activities typically begin on Saturday morning, with business meetings, professional sessions, a field tour to agricultural and urban sites, contests, and social events. Student events culminate in an awards program, business and planning meeting, and a farewell event on Monday evening. The scheduling of student events over an extended weekend minimizes time away from classes for students and faculty advisors and enables students to take advantage of lower airfares or ground travel to observe agricultural regions and soils in route to the meeting site. In 2006, 613 students were members of the

Table 2. History of student memberships and agronomy clubs.

Year	Number Chapters	Number Students	Notes and comments
1932			students first met in Chicago
1934	6		student chapters recognized
1936	13		representatives of tudent sections meeting annually
1942	23	500	membership certificates issued to students by ASA
1943–1945			student programs discontinued due to World War II
1947	17	592	temporary student officers named to rejuvenate activity
1950	30	1287	met with ASA away from Chicago for first time
1957	45	1287	
1964			graduate student membership made available in ASA
1975			membership opened for clubs from two-year institutions
1977	69		
1985	66	1200	
1995	56	1070	several existing clubs failed to pay dues
2006	69	613	women are the majority of annual meeting student attendees

A student proudly exhibits her poster during the Annual Meeting.

69 campus clubs and 527 of them also held a Society membership. The 2006 annual meeting was attended by 251 students. Six 1890 land grant universities had campus clubs.

PROGRAMS FOR CAREER ENHANCEMENT

ASA has a strong history of supporting student growth and professional development for both undergraduate and graduate students. In the past decade, ASA members have organized and presented programs to enable domestic and international students to gain perspectives and experiences for their post graduate career. From 20 to 60 students participate in one or more of the programs offered. The programs help students gain experience in job search and career development. For example, the Graduate School Workshop provides exploratory discussions for undergraduates on key considerations for graduate school, selecting a major professor and research program, performance expectations, assistantships and financial support, and travel opportunities. A mock interview program enables students to experience the expectations and questions of a corporate or graduate school interviewer and receive feedback on their strengths and improvement opportunities for future interviews. Through the Mentoring Program, Society members meet with individual students to discuss professional opportunities and challenges in the workplace. Other programs address job search strategies, writing manuscripts for publication, and grant writing skills. The ASA–CSSA–SSSA Career Placement Center provides a web-based listing of positions available and positions sought, helping students to search for graduate programs or career positions. At the annual meeting, students may interview for jobs or meet with representatives from universities to discuss graduate programs. Job positions are posted at the annual meeting, enabling graduate students and others to interview for public and private sector positions.

STUDENT CONTESTS AND RECOGNITIONS

Contests comprise a major part of the student program at national meetings, having evolved since the first grain judging in 1923 (Table 3). Rules and procedures have been revised based on experiences and contest goals. Today, officers of the student organization select a student member to lead each contest. Subcommittees for each contest help ensure continuity. Rules and procedures are passed along each year to an incoming chairperson. Most commonly, each campus club designates their best person or team to participate. Some contests are conducted at revolving locations, such as the soils contest, which rotates among regions. Most contests are held in conjunction with the national meeting, where club members vigorously compete. Some contests, such as crops judging, have a long history, whereas the national club poster presentation contest began in 2002. Some history and key features are summarized in Table 3.

Early contests were closely patterned after livestock judging contests and were instrumental in the formation of many agronomy clubs in the Midwest. In 1923, without ASA involvement, seven teams from universities in the Midwest participated in the first intercollegiate grain judging contest, which became the collegiate crops contest in subsequent years. The event was held in conjunction with the Hay and Grain Show of the International Livestock Exhibition in Chicago. With the exceptions of 1929, due the Depression, and the World War II years (1942–1946), this contest has been held annually. A similar contest was also held in 1929 in Kansas City, but later events were held in Kansas City and Chicago in the same week, typically with seven to twelve teams from the same institutions participating at both locations. Elling (1981) and later Posler and Higgs (2003) summarized details on college participation and placing of teams and individuals. Over the decades, at least 45 different schools participated, with 14 participating

Table 3. History of student contests and participation.

Contest	Starting year	Participation at national level	Participation trends	Other
Crops Judging	1923	7–14 teams	none apparent	
Essay/Manuscript	1933/1982	10–40 individual participants 5–20 clubs	highly variable	
Achievement Award, Club	1952	4–15 clubs	downward	last award in 2004
Soil Judging	1961	10–26 teams	upward	
Speech	1961	10–20 individual participants	none apparent	
Photography/Slide/Visual Presentation	1970/1990/1999	4–16 individual participants	variable, none apparent	
Research Symposium	1997	10–32 individual participants	upward	first papers in 1971 first poster in 2000
National Club Poster Presentation	2002	11–16 clubs	none apparent	replaces club achievement award

in 20 or more contests. Teams from Texas Tech University won 14 consecutive annual contests in Kansas City from 1958 to 1971. Ayers (1953), long-time coach of teams from Texas Tech University, reflected on the values and benefits of crops judging as an educational tool in agronomic educational programs. Crop judging, the most enduring of all student agronomy contests, has benefited hundreds of students over the years. Many alumni of crop contests pursued graduate study and later obtained positions of prominence in the agronomic sciences (Posler and Higgs, 2003).

An essay contest was initiated in 1933, and by 1937, 35 students participated, perhaps stimulated by cash awards provided by the Chicago Board of Trade. During its history, the essay contest experienced high fluctuations in participant numbers. As a result, in 1951, the contest was changed from writing a technical paper on a specified topic to preparing a semi-technical popular article suitable for ASA's *Crops and Soils* magazine. With the change in requirements, more papers were submitted. Winning essays were published, and authors received $25 and a year's subscription. In 1956, 29 essays were submitted from the 117 essays prepared in 16 campus clubs. When *Crops and Soils* magazine was discontinued in 1987 (reintroduced in Spring 2007 for certified crop advisors, extension scientists, sustaining members, and others), attention shifted to the oral presentation of research papers by undergraduates at the meetings. The essay contest transitioned into a manuscript contest, with the ASF Darrel Metcalfe Fund today providing cash awards to the winners. Annual participation in the manuscript contest is highly variable. In many respects, the original essay contest was the forerunner to other contests, including the manuscript preparation, visual presentation, speech, and research symposium contests that are now major contests for students at annual meetings.

An agronomy club achievement award was established in 1952 to recognize a campus club. Evaluation was based on club activities, membership, participation in contests, and other components. In early years the American Plant Food Institute provided a $100 cash award and trophy. Since participation in this contest was highly variable, it was discontinued in 1999. It was reinstated in 2001 but only attracted five entries. The award was last made in 2004. A national club poster presentation contest, initiated in 2002, now focuses on club programs and outreach, and has increased student interest in club competition.

Regional soils contests began in 1958, and by 1961 the first national contest was held. A SASES committee of ASA and SSSA members coordinates this contest at regional and national levels, and the two societies share equally in providing financial assistance to the clubs that host contests. Many rule changes over the years have accommodated new developments in soil taxonomy, scoring requirements, and greater consistency in evaluations and regional contests. Today, team placement in a regional contest is required for participating at the national level. National contests are rotated among regions where students from a host school assist with the contest but do not compete. As the soil judging contest became more popular, more schools participated in regional contests. More students participate in soil judging contests each year than in any other ASA student contest. Each year, with 150 or more students in the regional contests and 80 to 100 students on 20 to 25 teams at the national level, some 250 to 300 students participate in soils contests each year in the United States. Many current professors and leaders in agronomy participated in crops or soils contests, which foster concentrated study, dedication to purpose, attention to detail, and the ability to exercise good judgment.

The speech contest, first held at the national meeting in 1961, is conducted during the student program at the annual meetings. This contest, in which participants must have won a local contest, usually attracts 15 to 20 entrants. The contest typically consists of two concurrent preliminary sessions with a final session where the top three speakers

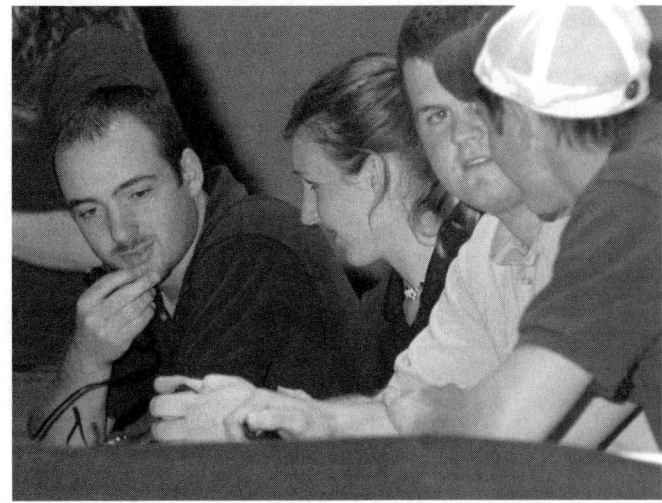

The student quiz bowl is a favorite event at the Annual Meetings, known for its friendly, but fierce competition.

from each preliminary event compete for final standings. On the day of the contest, speakers draw a card and select one of three topics for their speech. Contest rules include provisions on preparation, length of the presentation, and evaluation criteria used by judges. Participants range from freshmen and first time attendees to experienced seniors and accomplished speakers. Sessions are well attended by fellow students, who recognize that the speech contest is an important activity to enhance future success as an agronomist.

The photography/slide presentation/visual presentation contest has changed as visualization technologies have evolved, enabling student participants to gain experience with contemporary communication methods. As an offshoot from the essay contest, the first photography contest was initiated in 1970 to challenge participants to convey an agronomic message using photographs. The early emphasis was on the creative ability of the photographer to capture a message in pictures with explanatory captions. The photography contest shifted in 1990 to become the slide presentation contest, which involved multiple 35-mm slide images. In 1999, with the changes in visualization technology, this event advanced to become the visual presentation contest, utilizing digital technology, such as Microsoft Power Point. The contest involves preparation of 6 to 16 images and limited text to present a story or an overview of new technology.

The student research symposium, consisting of oral and poster presentations, is of particular significance due to the recent increased involvement of undergraduates in research. Increasingly, students are reporting research results in oral presentations, posters, and manuscripts from their actual work in research programs. Many of the presentations are part of the general ASA programs and are presented within the various sections of ASA, CSSA, and SSSA. However, undergraduate student research is also presented in designated oral and poster contests. Between 2001 and 2006 oral papers increased from 10 to 18 and student posters increased from 1 to 14, with a notable shift to more science-based titles.

Undergraduate students first presented research papers as a part of the program at the annual meetings in 1971. The research symposium contest actually began in 1997, resulting in the desired increase in number of papers presented by undergraduates. Initially, all presentations were oral, but in 2001, one poster and ten oral papers appeared in the program. Both types of presentations yield cash awards for the top three presenters. Participation has increased dramatically in the past five years to 14 oral and 18 poster presentations in the 2006 program. While winning entries normally report research efforts, 65% of the scorecard involves the effectiveness of the presentation, its organization and quality of visual aids. Attesting to the scientific quality and contemporary content, many professional agronomists attend the oral and poster sessions because the student presenters are frequently working in cutting edge science and may be prospective graduate students.

Since 2002 when the national club poster presentation contest was initiated, 13 to 16 campus clubs have presented posters at each annual meeting. This event enables clubs to showcase programs, highlight accomplishments, and exchange ideas with other clubs.

An addition to the national student program in 1984 was an evening quiz bowl. This friendly, enthusiastic competition involves teams of four students, frequently with ad hoc members representing one or more campus clubs. Much like a single elimination tournament event, it challenges two teams at a time to answer specific questions on all aspects of agronomy, soils, pests, and production practices. The quiz bowl is genuinely entertaining fellowship for teams, cheering sections, and audience members, while being educational and challenging even for professional members and judges. While no formal prize or presentation is provided, the winners enjoy "bragging rights" and, since 2004, take home a traveling trophy.

ASA began an awards program in 1957 to recognize an outstanding senior from each school. In the first year, 28 institutions participated and grew to attract selections from 40 to 50 institutions in recent years. The title was modified to "outstanding student" to accommodate nominees from 2-yr institutions, some of which have had recognized student chapters since 1975. The mid-1970s also saw a marked increase in the number of women recognized, with 12 of 40 in 2006 and 13 of 42 in 2007 outstanding students being women.

SUMMARY

The early origins and history of agronomic education were documented in committee reports in JASA. The ASA existed as a professional organization for more than a decade before significant attention was devoted to education endeavors at meetings or in journal articles. Perhaps ASA President Jardine established the initial expectations in 1917 for Society members by highlighting the need for well-educated agronomists for the future. Undergraduate involvements with ASA originated through crops judging contests held annually in Chicago and Kansas City, eventually meeting at ASA annual meeting sites. With only a decade of organized activity before World War II, a restart was well underway by 1950, and many educators and students were active in the Resident Education Division in the last half of the 20th century. They paved the way for expansive development of activity in resident education and student activities.

The JAE was established in 1972 and expanded in 1992 to become the JNRLSE and now attracts papers from other disciplines also interested in enhancing undergraduate education. In Resident Education, the number of published papers, as well as oral and poster papers presented at meetings, has increased dramatically since 1972. Topics such as outcomes assessment, curriculum, interactive learning, instructional innovations, and teaching resources have been featured in recent years. Through sustained support from ASA, JNRLSE provides a well-recognized venue for communication of scholarly activity in agronomic and life sciences education. Efforts to reach teachers of K–12 students have been expanded through development and publication of materials, as well as special programs at ASA meetings.

The student program at the national meeting has expanded from its original one-half day into a concentrated three-day event that includes business meetings, contests, tours, and social events. Career enhancement programs for students include graduate school and mock interview workshops, graduate student and internship interviews, and job opportunity sessions. More than 200 of the 600 students in the 69 affiliated campus clubs attend the annual meetings of the Societies and participate in SASES, their student organization.

Meetings of professional organizations have a less tangible, but highly significant impact on the professional lives of participating teachers and students. They learn and teach by interacting with their peers, thereby changing attitudes, acquiring skills, and expanding knowledge.

We wonder what the future holds for agronomic education, but the history of the Societies is encouraging in that the challenges for resident education in the first century have been met. ASA enters its second century from a position of strength in serving educators and students. The acquisition of knowledge, fueled by the ever-increasing need for production of food and fiber for more people on less land, and almost unbelievably rapid technological developments mean that change will be the order of the day. A solid foundation exists for resident education and student activities within ASA, CSSA, and SSSA.

Acknowledgments

Appreciation is expressed to Leann Malison of the Societies staff, who provided resource materials, reviewed the manuscript and answered many questions, and to Jim McKenna, Lowell Moser, and Gary Posler, who very graciously reviewed drafts and provided many good suggestions, corrected errors, and challenged us to find answers for questions that came to mind, and to others who responded to our queries.

REFERENCES

Ayers, C. 1953. Crops judging, grading, and identification contests. Agron. J. 45:565–566.

Brown, P.E. 1916. Soils courses at the Iowa State College. J. Am. Soc. Agron. 8:42.

Collins, M.E., and J. Pesek. 1983. Women in agricultural sciences. J. Agron. Educ. 12:87–92.

Elling, L. J. 1981. The intercollegiate crops contest: 1923–1978. J. Agron. Educ.10:5–13.

Ernst, S., and J.G. Graveel. 1996. K-16 science: Welcome to the K-16 classroom! J. Nat. Resour. Life Sci. Educ. 25:105.

Graveel, J.G. 1992. Journal changes title and scope. J. Nat. Resour. Life Sci. Educ. 21:1.

Jardine, W.M. 1917. The agronomist of the future. J. Am. Soc. Agron. 9:385–392.

Karraker, P.E. 1920. Notes on a conference on elementary soil teaching held at Lexington, KY, June, 1920. Soil Sci. 10:247.

Lyon, T.L., and H.O. Buckman. 1922. The nature and properties of soils. Macmillan, New York.

Metcalfe, D.S. 1958. The national student activities section: Past, present, and future. Agron. J. 50:488–490.

Metcalfe, D.S. 1962. Student activities sponsored by the Society. *In* H.H. Laude (Historian) History of ASA, 1907–1957. Agron. J. 54:64–66.

Pohlman, G.G. 1962. Divisions of the Society. *In* H.H. Laude (Historian) History of ASA, 1907–1957. Agron. J. 54:62–64.

Posler, G.L., and R.L. Higgs. 2003. The collegiate crops contests 1923–2003. K-State Printing Services, Manhattan, KS.

Extension Education

The American Society of Agronomy and Development of Agronomic Extension in the United States

Dwayne A. Rohweder
Emeritus Professor of Agronomy Extension
University of Wisconsin-Madison

The late 18th century marked the beginning of a general improvement in farming. Farmers began communicating more about new farming methods and new lands for settlement. Farmers began to explore new crops, new methods of farming, and new lands. The extension agent idea may not be as new as once thought. President C.W. Warburton (1925) in his presidential address reported that, as early as 1719, colonial documents expressed the need for special instruction in hemp raising, harvest, and manufacture. As soils "wore-out" from poor management, farmers left the depleted soils of the colonies behind for new land to the west. Early agricultural societies encouraged information sharing on new farming methods. Scientific discovery, new inventions, and increasing awareness of this information by people brought a new direction to agriculture.

Eighteen hundred sixty-two was a monumental year. The United States Department of Agriculture (USDA) was established. The Homestead Act was passed, spurring even more westward expansion and the development of the family farm. The Morrill Land Grant College Act, supporting the formation of colleges of agriculture and mechanic arts in every state, resulted in a great expansion of formal agricultural education. The Connecticut Agricultural Experiment Station was the first experiment station established in the United States in 1875, followed by passage of the Hatch Act in 1887 extending this opportunity to every state in the Union. By this time, the technology advances begun during the Industrial Revolution were well on their way to revolutionizing farming methods.

Many farmers in the United States were more focused on cultivating new western lands than on considering new ideas or improved methods. Farmers, and even scientists of the time, lacked knowledge of the best management of soils and plants, and yields after a first crop often dropped dramatically. One positive result came from the misfortunes of the wheat growers. Farmers began to study their own problems and began searching for alternatives.

In the waning years of the 19th century, scientists in colleges of agriculture began conducting outreach programs to farmers. Between 1880 and 1892, Farmers Institutes, or their equivalent, had been organized and were held in 26 states on a near permanent basis. At the same time, farm short courses for farm boys opened during winter months. State fairs and boys and girls clubs became new sources of agricultural education.

TWENTIETH CENTURY

The first recorded formal extension activity occurred in Hull, Sioux County, Iowa on February 18, 1903. This effort was followed in 1904 when Professor Seaman A. Knapp pioneered the "agricultural representatives" program in the southern USA as a means of fighting the boll weevil scourge on cotton. Farmers from the Sioux County Farmers Institute traveled to Iowa State College and enlisted the expertise of Professor P.G. Holden, Head of the Agronomy Department, to travel to Hull and teach Sioux County farmers how to improve corn production. A farmer

The author, D.A. Rohweder, Extension Agronomist, lectures on producing high quality alfalfa using new germplam. 1982.

Corresponding author
Dwayne A Rohweder, Emeritus Professor of Agronomy Extension, University of Wisconsin, Madison, WI (email: darohwed@wisc.edu).

Left: A demonstration on the topic of nutrient runoff.
Section page: An extension training session on forage establishment.

inquired whether agricultural experiments performed on the experiment station at Ames, were valid 100 miles away in northwestern Iowa. The first work undertaken was improvement of farmers' seed corn. The effort to improve corn started a movement that spread over Iowa and the Corn Belt. Analagous efforts began with other crops and in other regions. This general idea for extension outreach was not new (Apps, 2002). Agricultural exhibit or corn trains were an early innovation. Professor Holden conducted his corn trains in Iowa from 1904 to 1906. Wisconsin College of Agriculture's first project train started in 1904, with interest reaching a peak between 1912 and 1917. At the same time wheat and lime trains were conducted in Kansas.

First County Agricultural Representatives

W.C. Stalling, hired on November 12, 1906 in Smith County, Texas, was the first county extension agent employed in the United States. Iowa appropriated money for extension in 1906, and Wisconsin passed legislation appropriating $30,000 annually in 1908 (Schwieder, 1993). In 1912, the first county agricultural agent appointed in Iowa was M.L. Mosher in Clinton county. He was a charter member of the American Society of Agronomy and also was the last living charter member of ASA when he died in 1982. The first county agent appointed in Wisconsin was E.L. Luther in Oneida County. Other states also began hiring agents.

Early Involvement of ASA in Extension Development

In 1908, at the ASA meeting in Ithaca, NY, A.M. Ten Eyck (Kansas State), an ASA charter member and its third president, stated:

> The breeding of crops is a part of our agriculture, which until recent years has been largely neglected. Ten years ago very few experiment stations were doing any work in plant breeding. Today breeding of crops is the most popular work in agronomy, and almost every experiment station is carrying on some of this work. Several stations are not only breeding crops, but are producing and distributing among farmers considerable quantities of seed of improved varieties.
>
> (Ten Eyck, 1910)

In 1909, Professor George N. Coffey (Ohio State University and second president of ASA) stated:

> Probably no line of agricultural work...has had more rapid development and extension within the last decade than the classifying and mapping of soils. The work is changing attitudes of investigators toward many problems relating to the soil.... recognizing that soils possess an individuality as well as do plants and animals.
>
> (Coffey, 1911)

This work permitted extension and research workers to individualize recommendations to individual soil groups and to a specific farm.

Subsequently, in his presidential address in 1912, ASA President R.W. Thatcher (Washington Agricultural Experiment Station) recognized that:

> The greatest single problem confronting administrators in agricultural institutions is the proper distribution of funds and efforts of staff between research, extension, and demonstration fields. Taxpayers are demanding benefits. Associations, societies and companies having interest in the improvement of agriculture and food supply problems demand that land grant colleges provide more information, inspiration, and intelligence in methods of general farm practices.
>
> (Thatcher, 1912)

In his 1913 presidential address, ASA President L.A. Clinton stated:

> The greatest discovery in agriculture in recent times has been the farmer and his farm. In extension work more and more we are utilizing farmers to lead their fellow citizens to adopt better practices in agronomy. The development of separate extension staffs and especially the employment of extension specialists has done much to give continuity to research efforts.
>
> (Clinton, 1913)

The same year, A.M. Ten Eyck (1913), then at the Extension Department, Iowa State College, using ideas previously cited, published "A New Line of Agricultural Extension in Iowa." He said:

> The work is the investigation of local farm problems to determine and authorize a proper working system of soil management and farm practice for each farm or locality. An expert will propose a practical working plan regarding treatment of soil in different fields, rotation of crops, and other items which should be put into practice to make that particular farming enterprise more profitable and permanent. In his judgment, this direct application of science to the business and practice of farming by an agricultural extension service is of far greater importance and of more lasting value than platform speaking, bulletins and circulars which, up to this time, has constituted the larger part of the service given to farmers.
>
> (Ten Eyck, 1913)

ASA leaders were at the forefront of the development of the extension program in agronomy, as well as the extension program in the United States. They recognized the impact resulting from disseminating current and relevant research information to farm people.

Extension Gets Its Charter

In 1914, Congress passed the Smith–Lever Act providing for Cooperative Extension in all states. The passage of this act launched an era of progress in bringing agricultural research to the rural community. There was concern that the extension movement may have a negative impact on investigational agencies. With time the expansion of extension leads to a growth in research.

Trial by Emergency—1917–1925

With the advent of World War I, extension focused attention on providing food for the war effort along with feeding the American population—"Food Will Win The War." The National emergency ultimately highlighted the great importance of the Cooperative Extension Service in providing farmers with information on the best farming methods at that time. During the early 1920s, considerable thought was given to

A Wisconsin corn train by Professor P.G. Holden in about 1904. Photo courtesy of the University of Wisconsin-Madison Department of Agronomy.

coordination of the research and extension functions in agronomy to increase the efficiency of agricultural research and its application through extension. Of interest was that one-fourth of the program at the 16th annual meeting of the ASA was devoted to a discussion of extension work in agronomy.

In his presidential address, ASA President C.W. Warburton (1925), USDA Director of Extension, reported that the growth of agronomic science and membership since the founding of the American Agronomy Society in 1907 has been extremely rapid. Extension work had been organized, with county agricultural agents being employed in the North and South. Warburton noted that previously the researcher had to depend largely on his own efforts to distribute his work to the public, and "this he often was not well fitted to do," and the time needed for preparation, travel and presentation interfered with his work. With the passing of Farmers' Institutes, growth of the modern extension service, and the fact that extension agronomists totaled 109 in over 30 states, the extension agronomist was able to provide the contact between researcher and county agricultural agent and ultimately the farm family. Agronomists to date had been required to teach, among other things, general crops, soils, farm management, farm machinery, and meteorology. Gradual elimination of these diverse lines would improve the efficiency of both the research and extension agronomist. Questions often raised were: "What proportion of the facts by research is usable in extension and what proportion of the practices presented by extension are adopted by the farmer?"

Warburton (1925) also noted that departments debated employing part-time or full-time extension agronomists. Some stated that extension agronomists conducting their own research had a more intimate understanding of the body of knowledge; however it was difficult to find men adapted to both professions and difficult to administer. It was suggested that a better route was to have full-time extension personnel housed with their research and teaching colleagues for a close dialog of ideas. Warburton asserted that research and extension workers should be aware of and have a working knowledge of the scientific method in action, and that a thorough knowledge of practical farm affairs and an underlying sympathy for the problems of farm life, best learned in early boyhood, also are useful attributes for extension agronomists.

Many early agronomists in ASA were active in outreach activities along with research and teaching in their respective states. Some examples were Professors Willett M. Hayes (Minnesota) and R.A. Moore (Wisconsin) (Moore, 1898), who conducted early research and demonstrations on plant breeding. After three years of grain selection and improvement, Professor R.A. Moore at the University of Wisconsin enlisted the assistance of short course students and others to test and demonstrate new varieties around the state. In 1901, he called the students together and formed the Wisconsin Agricultural Experiment Association, an association that has remained a major vehicle in crops extension. In 1919, with thirteen state associations and Canada, he formed the International Crop Improvement Association and was elected its first president in 1920 (Rohweder et al., 2003).

Professor L.F. Graber (Wisconsin) in the early 1900s was a pioneer in the development of pasture renovation and promoting the culture of productive and persistent alfalfa in the North Central Region. Agronomists in the northeastern states (John Baylor, personal communication, 2006) were active in working in corn and pasture improvement, general crop fertility research and demonstrations. Professor T.A. Kiesselbach in Nebraska in 1915 summarized early soybean tests, stating that it was the most practical annual legume for cattle, hog, or sheep hay. His early work also included corn improvement and summarized climatic factors as they affected growing conditions and the growth and yield of corn (Frolik and Graham, 1987). Examples of extension topics reported at ASA meetings from 1918 to 1929 included U.S. Crop Centers, effects of lime and manure, lime surveys in Illinois, value of outlying experiment fields, utilizing legumes in rotations in the Northeast and Midwest, and the Massachusetts Better Forage Program for dairy.

Bracing Against Adversity—1920–1930s

Crushing agricultural deflation during the 1920s preceded the Depression of the 1930s. Prosperity collapsed nearly overnight. Hard times soon convinced state experiment station administrators of shortcomings in their research, with too little time spent on economic and social issues. Many papers given during the 1930s were devoted to defining the job of the extension agronomist, the relationship with research colleagues, and the role of science in developing action plans. Extension projects reflected the condition and needs of a post-war agriculture. Discussions at ASA meetings were devoted to evaluating teaching methods for several pressing crop and soil programs. Newer extension teaching methods had largely substituted for the old Farmer Institute and Extension Schools. Teaching aids such as news articles, radio programs, field demonstrations, talks illustrated with lantern slides, and bulletins and circulars were explored. Extension also returned to the practical and convincing field demonstrations that so often led to general farm acceptance (Graber, 1926).

A new concept in the role of the extension agronomist also was explored. In addition to bringing research information to the farmer, just as important was bringing back ideas and questions to the researcher for further study. Researchers often found their best leads came from farmers and their experience. J.S. Owens (1934) in Connecticut stated:

> Early leaders started with problems closely associated with agricultural practice. Contributions of soils are of concern in all crop production. Likewise, plant and chemical developments outside the narrow field of agronomy are seemingly increasing at a geometric rate and may mean more to farm crop production than those made by the more strictly farm crop research.

R.D. Lewis (Ohio) in 1937 (Lewis, 1938), quoted Pasteur's statement: "What really leads us forward is a few scientific discoveries and their applications." Lewis recognized the tremendous responsibilities that confront research and particularly extension in the correlation of knowledge and activities from all fields that affect the farm family, its home, and its enterprises. Lewis went on to describe how the function of the extension agronomist was going through transition. The extension agronomist was the interpreter, the teacher, the person to correlate those facts and processes that may be incorporated into scientific systems of soil management and crop production. He noted that the rapid development and creation of new materials meant that the extension agronomist must concern himself with methods of transmitting these research results quickly, effectively, and properly balanced to fit into agricultural problems of many people. Increasingly research, as well as extension, is exploring "the effects of science and its applications upon the individual and his society." The extension agronomist "must realize the probability that action will be taken on his recommendations" (Lewis, 1938).

During the latter years of the 1930s, agronomy extension programs also were being correlated with and related to programs of new federal agencies, including the Agricultural Adjustment Administration (AAA), Civilian Conservation Corps (CCC), Farm Security Administration (FSA), and Soil Conservation Service (SCS).

Topics important to Extension discussed at ASA meetings during the 1930s were the selection of drought tolerant legumes, winter pastures, hybrid vs. open pollinated corn, mixing wheat and rye, quality tobacco, rapid soil tests, cutting frequency of alfalfa, and foundation seed stocks.

World War II and its Aftermath—1940–1950

For the second time in a generation, farmers were asked to produce food, feed, and fiber for a nation at war—"Food Will Win The War and

Write the Peace" (i.e., the Marshall Plan). The years from 1941 to 1946 were difficult for farm families, but farmers came through with unparalleled production. In addition to traditional crops for each region, several critical crops shared the spotlight for the war effort, such as soybeans for grain, dry and canning peas, dry beans, flax, hemp, and potatoes. Economical production and quality products became essential parts of every program dealing with farm and factory development. The greatest opportunity for the farmer still rested in the improvement of individual efficiency, and extension worked to help farmer's realize this potential.

The Veterans' training bill at the close of WW II provided an opportunity for returning veterans to attend universities, colleges, and technical schools, increasing the numbers of person educated in agronomy. Many of these veterans entered extension as extension specialists and county faculty, permitting a marked increase in extension programming.

At the 1947 ASA meetings, Professor H.B. Cheney (1947) reported that the extension program on fertilizers in Iowa was developed to meet a recognized need of furnishing Iowa farm people with more accurate information on the proper use of fertilizers to fit their particular soil and crop conditions. Recommendations were based on extensive research in the major soil areas of the state. He noted that result demonstrations did much to stimulate interest. The program was based on meeting an important need of farm people and supplying the information through research. Extension methods utilized to provide the assistance to farmers were soil tests, leader training, farmer meetings, tours, and publicity. It was a continuous program over years.

The War Boom Collapses

The prosperity of the war years continued into the early 1950s. Farm families were living a better life. Science was making their jobs easier, and they were enjoying most of the comforts of urban life. However, by 1953 surplus production caused lower farm prices, higher costs, a drop in livestock prices. After making an all-out production effort for years, many farmers began to adjust their operations, thinking in terms of increased efficiency. Attention was not only paid to crops and fertilizers, but also pests and diseases.

At the same time, extension work was flourishing, delivering years of research by USDA, state experiment stations, and private firms to rural people. Farmers were looking at all aspects of the farm enterprise. Because of low livestock prices, farmers sold their herds and adopted the "Corn–Soybeans–Miami" rotation, giving many farmers their first vacation ever. Continuous row-cropping, with its associated soil compaction problems, tillage options, fertilizer rates, and pest concerns, created new problems to be addressed by agronomy research and extension personnel.

In addition to traditional agronomic activities, extension agronomists assisted in Farm and Home Development programs. Economic and social issues also influenced recommendations, and women became more frequent audiences. During the late 1950s, extension agronomists began to wonder why farm people were not adopting recommendations as expected. Thus, they shifted meeting format on corn, soybean, and forage production from short production and management topic meetings to three-day in-depth schools. Recommendation adoption increased because attendees better understood why they should consider changing their practices.

Professor J.C. Lowery (1957) (Alabama), in his address at the 50th anniversary meeting of ASA reflected on "The Extension Agronomist—Past, Present, and Future." He said, "The extension agronomist occupies an important position in the Land Grant College System. His responsibilities are great, his opportunities almost unlimited, and his field a broad one. He cannot be simply a teacher of subject matter furnished by the resident teaching staff. He must analyze the needs of his state and individual counties and develop a comprehensive program to meet these needs." Lowery noted that the extension agent of this time must spend considerable time in study of research from his own experiment station and other stations. His responsibilities reach all farms. He also works with business and industry—seed companies, lime and fertilizer industries, cooperatives, and marketing companies.

Harold E. Myers (1958), Dean of Agriculture at the University of Arizona, at the 1958 ASA annual meeting discussed coordination of research, teaching, and extension and noted that several organizational patterns have developed over the years in the Land Grant College system. In some instances the Director was directly responsible to the President of the university. In some, the Director was reported to the Dean of Agriculture who was responsible for an integrated agricultural program. Through the years, more and more institutions moved toward this pattern of integration.

A Catalyst For Change

By the close of the 1950s, agriculture was poised for a major change. In 1959, a symposium entitled "Adjustments in Agriculture" was organized by the extension education subdivision. Dr. Earl L. Butz (1959), Dean of Agriculture at Purdue University stated that "This is the age of science and technology, based on research. The frontiers of the mind have replaced the frontiers of geography." Butz observed that the scientific frontier in America is barely scratched and has no effective limit; organized and imaginative research will push the scientific frontiers beyond the limits we scarcely dream of today. He said the area for expansion based on experimentation and innovation "staggers the imagination."

Dr. Marvin Anderson (1959), Associate Director Iowa State College Cooperative Extension Service, observed, "Change is the most significant feature characterizing American Agriculture in the last 50 years and particularly the last generation....The sobering evidence is that farmers have not shared proportionately well with the general economy in this progress." Anderson described how increased output is no longer the number one criteria, and management skills are increasingly complex and important. He urges extension to interrelate plant science to animal science and economic and social science, for a more coordinated effort.

Dr. Harold Jones, Director of Cooperative Extension, Kansas State University, spoke on the role of extension in the changing scene of agriculture (Jones, 1959). He described how farms are larger and farm populations smaller, and more farmers and their wives are working part-time off the farm. Efficiency of production has increased markedly. Capital in the form of machinery, fertilizers, and labor saving devices have been used to offset rising labor costs. He viewed the change as not a dwindling agriculture, but rather a strong agriculture as a segment of the national economy, if we consider the high proportion of non-farm people who are engaged in allied agricultural industries. He also described the increase in non-farm consulting by extension.

George Enfield (1959), USDA Agronomist, described the three main concerns of extension work: 1. Who is the audience? 2. What are the answers to the questions they may ask? 3. How can these skills and materials be transferred most effectively and efficiently? Enfield forecast fewer farmers and more suburbanites who garden for fun or recreation, while farms will be larger and more specialized. He said farmers that specialize and know more about their job are hungry for additional information, and more farmers will have had agricultural training. The divide between outstanding and average farmers will be greater. Farmers will be more aware of the need for keeping up with new ideas and know more about how to obtain needed information. Research results will be more fragmented, smaller, and written in technical jargon that will need translating and interpreting to make them meaningful for the extension worker. The specialist will find he will need to spend more time studying the fields related to his subject matter to best determine the trends in these fields and the effect on his area of responsibility. Specialists will find it essential to cooperate with specialists in other fields to avoid conflicts in recommendations.

Organization of the A-4 Agronomy Extension Division

The ASA extension programming was first separated out as a subsection of the provisional Agronomic Education Section XIII in 1947. The Agronomic Education Division became permanent in 1949 with Extension Participation as a subdivision. This organization continued through the 1950s. The symposium on the "Adjustments in Agriculture" listing many challenges to agriculture, agronomy, and implications for the extension specialist was accepted as a challenge by the officers of the Extension subdivision and set the stage for the marked increase in the number of papers by extension agronomists in the Society. Also in 1962, the agronomy extension education subdivision sponsored a symposium on the needs, challenges, and opportunities for agronomy (climatology, crops and soils) undergraduates in positions with Cooperative Extension, Soil Conservation Service (SCS), seed and fertilizer industries, and as field experiment station personnel. Membership in the subdivision also included industry agronomists and field station supervisors in anticipation of forming separate divisions.

In a 1962 letter to the ASA Board of Directors, Dr. Frank Schaller, Iowa State College Extension Agronomist and Board Representative stated that A-1b, Agronomy Extension Teaching subdivision, had reached a stage where a more formal organization separate from Division A-1 appeared necessary (1962 A-1b Minutes, ASA). This more formal organization would permit upgrading of division programs and wider participation by extension agronomists. The goal also was to provide an opportunity for extension agronomists to demonstrate how research findings are interpreted to action programs, that extension agronomists analyze broad problems, plan for the future, and truly play a major role in our country's land use and food production. He proposed Divisional status for Agronomic Extension Education.

Broad program topics reflecting the emerging interests in extension agronomists continued. In 1963, the Agronomic Extension Education Subdivision conducted a half-day session of invitational papers and two full day sessions joint with other divisions. Iowa State College Professors E.R. Duncan and R.H. Shaw (Duncan and Shaw, 1963) spoke on the importance of including agricultural climatology and weather as an integral part of management considerations.

In minutes of the Extension Education business meeting, November 19, 1963, Dr. Frank Schaller recorded a record attendance of 131 at the A-1b breakfast and business meeting. At that meeting extension subdivision members voted unanimously to request full divisional status from the ASA Board of Directors. The Extension Teaching Division A-4 was granted provisional status on November 19, 1964. Thus, both Divisions A-1 Resident Education and A-4 were on Provisional Status from 1965 to 1966. Ratification of proposed A-4 by-laws was approved at the 1966 annual meeting. Membership in A-4 was to include "University and Industrial Agronomists engaged in the dissemination of agronomic information on a non-resident instruction basis."

Additionally, considerable concern was expressed on the problem of obtaining publication of extension papers in *Agronomy Journal*. A committee of Dwayne A Rohweder (chair), Robert Dennis, Ken Morrison, and W.D. Pardee was appointed to prepare and submit a draft on publication concerns to the ASA Executive Committee.

On August 21, 1966, the provisional A-4 Agronomy Extension Education division requested that the ASA Board of Directors grant permanent divisional status in the American Society of Agronomy. Professor J.V. Baird wrote that "Extension Agronomists in ASA have functioned for many years as sub-division A-1b, Extension Teaching. Each year a well attended annual breakfast and business meeting have been held followed by a voluntary paper session. The group also has participated in invitational symposia....Extension agronomists are first and foremost agronomists. Some also are involved with research and resident teaching, yet we have a unique responsibility, that of interpreting and extending agronomic research information to farmers and related industry. Divisional status provides greater recognition of this important role and enhances the opportunity to meet our specific needs for professional improvement and to work more effectively with other segments of the Society." Permanent status for Division A-4 was granted in November 1967.

In 1970, it was moved that Divisions A-1 and A-4 cooperate with ASA in the initiation of a new journal for publication of papers dealing with agronomic education, the *Journal of Agronomic Education*.

A-4 Agronomy Extension Education Programs

Programs and symposia during the next four decades dealt with the changing needs of extension agronomists as they expressed the changing needs of both the rural and urban clientele in the United States and world wide. From 1964 to 2006, the A-4 Division conducted and/or cooperated in joint programming for nearly 500 sessions. More than 100 of the sessions dealt with papers on "outstanding programs that worked" conducted by extension agronomists in their states. The remaining 400 joint sessions or symposia were developed around "current topics on interest" to share changes within and without the profession that impact on their programming. More than 2500 papers were presented. During this time A-4 conducted joint programs with all of the A, C, and S divisions.

Crop Production and Management. Crop production technology was often the theme for papers of successful extension agronomy programs. In the 1960s and 1970s papers emphasized narrow row farming in corn and cotton production, meeting the challenge of the "variety development explosion" by commercial companies and universities, using plant analyses to determine problems in crop management, and development of 4-H projects to incorporate science into the project. Since the advent of continuous row cropping in the early 1950s, many questioned, "How High can Yields Go?" Professor D.A. Rohweder (A-4 Program Chair in 1966) organized a symposium, "Maximum Crop Yields—the Challenge," addressing topics of dry matter accumulation by plants, soil and environmental limitations, and fertilizer needed for maximum yields. One speaker addressed a case study on how a corn grower increased yield from 90 to 200 bushels per acre. Division S-8 sponsored an associated symposium entitled "High Yield Efforts—Fertilizer Management Practices with Soil Plant Implications." Papers from both symposia were included in the ASA Special Publication 9 sponsored by A-4, C-3, and S-8 (Rohweder and Younts, 1968). These sessions were followed by a 1970 symposium addressing "Moving Off the Yield Plateau." In 1971, the topic was "Maximum Economic Yield," and in 1978, "Moving Up the Yield Curve."

In the 1980s there was an explosion in crop production program topics. Early topics dealt with the impact of maximum yields on corn profitability. In 1987, the focus was "Maximum Economic Yield." Programs dealing with the increasing use of the soybean and using stages of growth in soybean management and in determining hail loss were frequent. Crops such as high oil corn, rice, sugarbeet, canola, cotton, triticale, and exotic vegetables along with low input farming were discussed. Preserving the environment was a frequent subject. Crop cultivars for specific uses and locations, nutrient and pest managements, double cropping, new crop management systems including cover crops, coping with the new genetic traits (e.g., glyphosate and corn borer tolerance), and precision farming were popular topics for programs and symposia. Sustainable agriculture practices were discussed, including alternate strategies for weed control in conservation tillage and in row and legume crops. Weed identification for professionals and farm people was discussed. Tolerance of legumes to corn herbicides was a topic discussed in the mid 1990s. Precision farming was the focus in the 1990s, with emphasis on the sciences dealing with row crop production, causes of yield variability, managing low producing areas in fields, compaction, and managing plant root systems for efficient crop production.

The decade of the 2000s included continued study on developing programs for change. The decade began with a concerted effort to improve field extension faculty and practicing agronomists' understanding of the genetic engineering process, biotechnology, and genetic modified organisms (GMOs) and their unique crop management considerations. Extension examined its role, the challenges and implications of educating about GMOs and seed arbitration because states were receiving many complaints about Roundup Ready (RR; Monsanto Technology LLC, St. Louis, MO) and Bt corn and cotton. However, one state reported on educating the public informing them that RR and Bt corn and cotton have saved producers millions in pesticide costs and reduced the potential for environmental contamination. Corn GMO hybrids must be compared using sister lines for adequate evaluation. Extension and industry are developing partnerships in training personnel. Extension agronomists also are forging effective partnerships to advance environmental education and sustainable agriculture. Other areas of discussion included integrating management and ecology in corn and cereal crop production, seed production in transgenic crops, and protecting the environment. Crop production papers focused on eliminating roadblocks or risks to production efficiency. The effect of irrigation on soybean production efficiency, abiotic and biotic influences on yield variability, Roundup Ready, soybean cyst nematode, iron chlorosis, and date of planting were all topics addressed that affected efficient production. The science of turfgrass management and production were topics of programming in extension sessions into the 2000s. Future turf care, roadside vegetation, salinity on turf culture, and bermudagrass were subjects of discussion.

Field Research Programs. Symposia were conducted reviewing the results obtained in long-term experiments conducted at agricultural experiment stations, including the Morrow plots in Illinois, the Sanborn plots in Missouri, 77 years of corn research in Nebraska, Alabama's old rotation experiment, and the oldest experiments in the South on cotton, and reference was made to work on the Iowa State rotation plots. The benefits of integrating extension and research, value of interactive research, on-farm research, and conducting performance tests were topics in the 1990s. Conducting adaptive and organic research were program topics in the 2000s.

Pastures and Forage Production. Much early emphasis on forages in the ASA addressed methods of improving pastures. Nonimproved pastures often were low in productivity due to lack of fertility, proper management, and lack of drought tolerance. Research in the early 1900s across the northern and central USA dealt with introducing drought-tolerant legumes into grass pastures and then introducing alfalfa and deep-rooted grasses into harvested forages to gain drought tolerance, longer stand life, quality forage, and higher yields. Outstanding extension programs dealing with forage crops across the United States were early paper topics (1925–1962). In the 1970s, pasture demonstrations again were addressed, as well as obtaining higher forage quality in species and varieties for use in balanced feeding rations. The 1980s included papers on grassland renovation, forage yield determination in mixed swards using a disk meter, top alfalfa growers programs, higher quality and feed analyses in balanced rations, near infrared reflectance spectroscopy in extension forage testing programs, and new national hay standards based on quality (Rohweder et al., 1978). Topics addressed in the 1990s dealt with quality hay auctions, certification of commercial forage testing laboratories, the PEAQ system for evaluating forage quality in the field, annual forages and alternate crops, forage establishment and grazing management techniques, effect of high and low traffic levels on alfalfa plant survival, grazing land conservation, a national alfalfa intensive training program, and a symposium on forage variety testing.

Papers in the 2000s continued the exploration of forage management for nutritional value and fertilizer nutrient management in forage–livestock and grazing land systems. Incorporating the science of ecology in forage crop management also was being emphasized.

Soil Morphology, Surveys, and Land Use. Soil mapping and land use was one of the earliest topics addressed in ASA programs. They have continued to be an important subject for field advisors into the 1980s. Programs adding to the understanding in this field were extension teaching in soil classification, physical properties of soils leading to high yields, agricultural land use planning, land use patterns, standardized soil survey information, soil programs from a realtor's point of view, soil maps, short course for county agricultural agents, landlord conferences, and small land holdings. Papers dealing with soil compaction were presented in the 1970s, leading to symposia on soil technology and soil compaction in the 1980s. Nonagricultural land use planning was an emerging program topic in the 1970s. Soil surveys led to papers on tax assessment and applying soil science to environmental health.

In the 1990s prescription farming focused on soil variability. Symposia on soils, environment and, water quality and integrated farming systems, including conservation tillage. Additional papers covered urban soil problems, such as urban forestry, soil conservation and restoration with the petroleum and mining industries, effective use of composts, manures, and organic materials in rural and urban settings. Programming in the 2000s concentrated on site-specific soil management and on urban soil problems.

Soil Fertility and Testing. Early extension related papers reported on the effective use of lime and manure (1923), rapid soil tests (1936), and establishing soil testing laboratories in the mid 1940s. Soil fertilization as a part of the war effort to obtain top yields along with balanced farming were discussed in the late 1940s. The A-4 programs in the 1960s, 1970s, and 1980s addressed development of soil tests, working with commercial soil testing laboratories, increasing fertilizer use, managing nitrate nitrogen, liming for profit, and education on crop responses to various fertilizer nutrients. Soil conditions and nutrient management, critical levels of nutrients in the soil and mineral composition of plants in relation to human and animal health, and changing recommendations to control pollution and improve water quality were subjects during the same period. Symposia were held on soil testing, sampling, yield correlation, calibration, and interpretation.

The decade of the 1990s featured soil fertility relative to maximum economic yield, nitrogen management practices to protect water quality, prescription farming, and integrated farming systems. Training on soil and water chemistry and nitrogen management to protect water quality and stewardship on the part of fertilizer dealers were topics. Toward the end of the decade increasing use of computers to modernize soil test recommendations to integrate use of manures, residues, poultry litter, fertilizers, in relation to soil test levels were discussed to improve efficiency of crop production as well as preserve the environment and water quality. Diagnostic tools were discussed as well as testing for phosphorus levels in soils. Nutrient management, use of the internet for recommendation dissemination, and regenerative agriculture were programs in the 2000s. Do-it-yourself comprehensive nutrient management plans workshops in one state helped producers understand why they needed to make adjustments. Other states adopted volunteer nutrient management programs. Determining total maximum daily load relative to N-load reduction in Southeast water industry was done in educational programs.

Extension Methodology. Extension teaching methodology gained more emphasis in the 1960s. For example, in 1967, "The Relation of Extension Activity and Applied Research" and a "Review of Outstanding Extension Programs" were the subjects of two half-day programs. With the latter program, states were asked to send copies of effective publications and slide sets to George Enfield, USDA Extension Director, for display at the 1967 annual meeting in Washington, DC. Agronomy training, exploring the potential for cooperation of extension and Industrial Agronomists, roles of Area Extension Agronomists, and Extension Agronomists with split appointments in applied research were the key subjects of programming in 1968. Visual aid development was addressed in 1969. Opportunities in programming with industry, roles of extension agronomists, teaching methodology and how to evaluate it, and using effective communication were addressed. Working with urban clientele presented new opportunities during the 1970s.

The 1980s began with programs that reached nontraditional audiences. Programs also were developed around teaching off-campus with resident faculty, interdisciplinary crop programs, using experimental designs in demonstration plots, regenerative and low-input agriculture, and sustainable agriculture. Programs in the 1990s dealt with incorporating more science into extension education programs. The decade began by featuring a roundtable evaluating case studies of "successful programs that worked" and discussing the changing needs of the extension agronomist. Techniques and issues of incorporating soil science, plant physiology and plant anatomy into crop and soil management practices was a feature. Topics included agronomic practices in a systems approach, techniques and issues of working with professional organizations, use of integrated pest management. Continuing emphasis was placed on conventional, low input, organic, sustainable, and multicropping farming systems. Agronomists discussed innovative techniques to enhance extension's role in disseminating data to additional clientele, including agribusiness, online audiences, crop consultants, as well as technology transfer through the media and broadening horizons in extension teaching. They discussed designing diverse futures in agronomic crops, soil and plant taxonomy, and developing centers of excellence targeting county agricultural agents, and dealers as well as the role of nongovernmental organizations functioning in the role of governmental organizations in technology generation.

International Agriculture. As extension agronomists became increasingly involved in International programs, A-4 cooperated in sponsoring programs featuring topics on establishing graduate programs and research programs in foreign universities to educate professionals to assist developing countries in improvement of heir food production and economic status, correction of slash and burn farming in the tropics, rice production, international agricultural trade, and global sustainability.

Agronomy Programming in the Electronic Age. Extension Specialists were early adopters of advances in electronic technology. The computer and software became a useful tool in extension programming over the years. The use of computer technology first appeared in A-4 programs in 1965 and 1969, concerning their use in data processing, computerized soil test reports, and as a tool for meeting the challenge of the variety explosion. In the 1970s, the computer was used with new tools and systems in interpreting soil maps by use of remote sensing of soil–water–plant relations. Subsequent to the decade of the 1970s, more than twenty symposia and paper sessions were conducted on the topics of computer software and the potential and use of computers in extension programs.

During the 1980s, sessions continued on integrating soil classification information into extension programs, and crop data modeling, as well as modeling in plant and soil systems, technology transfer into crop production and management systems, and aids for summarizing research. In the decade of the 1990s, the computer was used intensely to

Learning about new forage analyses during a Wisconsin Forage Council field day.

integrate climate, pH levels, tillage methods, manures, nitrogen, residues and fertilizer nutrients into more complex but efficient nutrient systems for crop production. Computers as technology in agricultural research and education and for modeling in plant and soil systems facilitated development of extensive and more detailed fertilizer recommendations, soil–water–and climate programming, crop management, on-farm testing programs and in computer based decision making. The computer was a tool in expert systems and a tool for crop management programs. The decade of the 2000s brought discussion on the use of fiber optic networks, satellites, distance education and multimedia as new tools. Remote sensing to measure spatial variability in fields to improve soil management also received considerable emphasis.

Practicing Agronomists. People interested in A-4, Extension Education, were significantly involved in developing the standards for ARCPACS and ICCA Certification. A workshop on obtaining certification and working with Certified Crop Advisers (CCAs) and ARCPACS certified consultants was conducted by A-4 and ASA officers. With the rapid increase of CCAs beginning in the 1990s and early 2000s, A-4 members were instrumental in developing the new A-9 Division, Professional Practitioners.

SUMMARY

In the modern Land Grant college system, Cooperative Extension was the third or last function created. Dr John Pesek, Emeritus Professor of Soils and Head of Agronomy at Iowa State University and past president of the American Society of Agronomy (personal communication, 2006) speculated, "If one considers the number of complaints from farmers during the mid to late 1800s and beyond, one might conclude that extension should have been the first function created because the lack of information available for farmers suffered the same shortage of organized and verified information faced by teachers of agriculture." From the beginning, officers of the American Society of Agronomy took an active role in the development of the A-4 Extension Education Division. The extension specialist has often taken the lead in advancing new technology. Sometimes, researchers and extension specialists have been criticized for not moving fast enough in providing information needed in this rapid moving society.

A review of A-4 programming shows that agronomy specialists have been leaders in the interpretation and dissemination of valid research information to solve problems of rural and urban clienteles. They have progressed from primarily crop production specialists to be members of interdisciplinary teams incorporating science with practice to improve understanding, acceptance, and ultimately adoption. They marched through the information and technology ages into the biological science age to assist society in solving complex problems dealing with both improving food and feed production and sustaining the environment and water quality. Extension agronomists will continue to be educators of the public with little knowledge of their relationship to and dependence upon plants, soil, and water, a public who has no idea of how their tastes and habits affect their basis for continued existence on the earth.

Today, extension agronomists are involved in biodiversity, biotechnology for energy production, biogenetics, global warming, globalization, sustaining the environment, and maintaining water quality, to name but a few. The extension agronomist today and in the future must be a top field agronomist. In addition to results from any of their own research, they also must be familiar with the work of other researchers to make sure that clientele obtain the complete message. To do this, agronomists are creating crop reporting networks, web sites, and using the internet to link agronomists in a particular discipline. The agronomist will use county extension faculty having advanced research degrees and CCAs to assist in obtaining information and disseminating that information in effective programs. Today, biogenetics, engineering, and the Evergreen Revolution involving scaled-up management of inputs on new technology will help solve hunger and poverty while sustaining the quality of the environment. They are opening new vistas for improving the nutrition and health of the world population and for development of alternative forms of energy. Extension agronomists, and their understanding of crops, soils, and climate, will be crucial in these national and international movements.

REFERENCES

Apps, J. 2002. The people came first—A history of Wisconsin Cooperative Extension. Wisconsin Epsilon Sigma Phi Alpha Sigma Chapter, Madison, WI.

Anderson, M.A. 1959. Adjustments in agriculture—A challenge to agronomy. I. Implications to agronomic education. Agron. J. 51:500–502.

Butz, E.L. 1959. Education—Our inexhaustible resource. Agron. J. 51:497–500.

Cheney, H.B. 1947. Extension program on fertilizers in Iowa. J. Am. Soc. Agron. 39:300–307.

Clinton, L.A. 1913. The agronomist in his relation to the farmer. Proc. Am. Soc. Agron. 5:193–202.

Coffey, G.M. 1911. The development of soil survey work in the United States with a brief reference to foreign countries. Proc. Am. Soc. Agron. 3:115–129.

Duncan, E.R., and R.H. Shaw. 1963. Agricultural climatology in an agronomy extension program. Agron. J. 55:299–302.

Enfield, G.H. 1959. Adjustments in agriculture—A challenge to agronomy. IV. Implications to extension service. Agron. J. 51:506–507.

Frolik, E.F., and R.J. Graham. 1987. College of Agric. of the Univ. of Nebraska-Lincoln—The first century. Board of Regents, Univ. of Nebraska.

Graber, L.F. 1926. Visual aids in extension work. J. Am. Soc. Agron. 18:26–29.

Jones, H.E. 1959. Adjustments in agriculture—A challenge to agronomy. III. Implications to extension education. Agron. J. 51:504–505.

Lewis, R.D. 1938. Integrating research and extension in agronomy. J. Am. Soc. Agron. 30:179–187.

Lowery, J.C. 1957. Extension agronomist—Past, present, and future. Agron. J. 49:645–646.

Moore, R.A. 1898. History of the Univ. Wisconsin-Madison, Agronomy Dep. *In* D. Rohweder et al. (ed.) Board of Regents, Univ. of Wisconsin, Madison.

Myers, H.E. 1958. Coordination of research, teaching, and extension. Agron. J. 50:486–488.

Owens, J.S. 1935. The interdependence of agronomic research and resident and extension teaching. J. Am. Soc. Agron. 27:413–416.

Rohweder, D.A., R.F Barnes, and N. Jorgenson. 1978. Proposed hay grading standards based on laboratory analyses for evaluating quality. J. Anim. Sci. 47:747–759.

Rohweder, D.A., D. Peterson, and J. Lauer. 2003. Univ. Wisconsin-Madison Agronomy Department—The first 100 years. Board of Regents, Univ. of Wisconsin, Madison.

Rohweder, D.A., and S.E. Younts. 1968. Maximum crop yields—The challenge. ASA Spec. Publ. 9. ASA, Madison, WI.

Schwieder, D. 1993. 75 Years of Service—Cooperative extension in Iowa. Iowa State Univ. Press, Ames, IA.

Ten Eyck, A.M. 1910. Breeding, multiplying, and distributing improved seed grain by the Kansas Experiment Station. Proc. Am. Soc. Agron. 1:70–82.

Ten Eyck, A.M. 1913. A new line of agricultural extension in Iowa. J. Am. Soc. Agron. 5:54–55.

Thatcher, R.W. 1912. The relation of research to demonstration work in agriculture. Proc. Am. Soc. Agron. 4:27–34.

Warburton, C.W. 1925. Taking agronomic research to the farmer. J. Am. Soc. Agron. 17:757–764.

ASA in the World

The International Dimension of the American Society of Agronomy: Historical Perspective, Issues, and Challenges

The task of chronicling the history of the international involvement of the American Society of Agronomy (ASA) for the past 100 years is a daunting one because of the many international aspects of the Society. One has to depend on the archives for the early years of ASA before the soils division became the Soil Science of America in 1936 and again when the crop science divisions became the Crop Science Society of America in 1955. In essence, ASA assumed all three Society functions from its inception in 1907 to 1936. Subsequently, despite separate structures, the Societies had much in common. We have attempted to deal with what was unique with respect to ASA and its international programs.

With brief allusions to the historical evolution of the Society, we teased out evidence of internationalism from the archived literature. The growth in numbers of international Society members and the production of publications of international interest documented the growing importance of international agriculture in the history of the Society. It was particularly relevant to highlight the driving force behind international agriculture—world hunger and the need to eliminate poverty from the globe. It was pertinent in this endeavor to highlight the Green Revolution and to feature some of the luminaries on the international scene, especially for wheat and rice.

The second half of the 20th century is notable for the various international institutions that joined hands with ASA in its noble role of making the world a better place for humanity. Particular attention was given to the division of ASA specifically designated to deal with international agriculture. As with any history, one has to focus on people, especially the ones that were recognized by society at large, and by ASA, for their outstanding contributions to the goals of the profession. Like all histories, it is a work in progress. The world is constantly changing; as some problems are solved, new ones need to be addressed. The world of tomorrow, with perhaps 10 billion people, will pose new challenges for ASA and its associated international division. Bright young and idealistic minds are needed now more than ever before to apply science to ensure that the world's population is fed and clothed, and that this can occur without loss of the resource base or damage to our global environment. It is our goal to inspire future generations to meet this noble challenge.

THE EARLY YEARS OF THE SOCIETY

Although the ASA was founded in the United States, its objective, "The increase and dissemination of knowledge concerning soils and crops and the conditions affecting them" (Laude et al., 1962), and its definition of agronomy, "a study of field crops and their relations to the environment" (Burton, 1963), are not at all confined to the United States. Clearly, the development of U.S. agronomy as a science was influenced by international agronomy. The ASA's first president, Mark Carleton, who traveled extensively to Eastern Europe and South Amer-

W.A. Payne
Professor of Crop Physiology
Texas Agricultural Experiment Station, Texas A&M University, Bushland

J. Ryan
Soil Fertility Specialist
International Center for Agricultural Research in the Dry Areas (ICARDA), Aleppo, Syria

Sustainably feeding and clothing the ever-growing world population is arguably the most inspiring and challenging work that ASA members have faced historically and into the next century.

Corresponding authors
W.A. Payne, Professor of Crop Physiology, Texas Agricultural Experiment Station, Texas A&M University System, 2301 Experiment Station Rd, Bushland, TX 79012, USA (w-payne@tamu.edu).

J. Ryan, Soil Fertility Specialist, International Center for Agricultural Research in the Dry Areas (ICARDA), P.O. Box 5666, Aleppo, Syria (j.ryan@cgiar.com).

Left: ASA's history reflects international cooperation and a vision for global solutions. Photos: NASA-Johnson Space Center and USDA-NRCS.

Section page: Southwest Arabian Peninsula, northern Somalia, and the Great Rift Valley of Ethiopia. Photo from Image Science and Analysis Laboratory, NASA-Johnson Space Center.

ica, explicitly recognized international contributions to U.S. agronomy, particularly those from Europe (Carleton, 1910). Half a century later, Bradfield (1957) again recognized international contributions to the success of U.S. agriculture:

> I would like to point out that other countries have made, through the years, many important contributions to American agriculture. Most of our improved breeds of livestock have been imported from Western Europe. Alfalfa came from the Near East; the soybean, the second most important crop in the corn belt, came from China. Many countries have contributed genes to many of the improved varieties of wheat grown in this country. Plant explorers are still scouring the earth for new crops.

International Membership

When ASA first met in Chicago during the winter of 1907, all 43 founding members were from the United States. Three years later, there were seven members from Canada and one from Mexico (ASA, 1911). By 1918, the Society's roster included 37 members from 11 countries outside the United States, including 21 from Canada, 5 from Cuba, and 2 each from India and South Africa. In 1953 (ASA, 1953), ASA had 169 members from 44 countries, including 52 from Canada, 10 from Mexico, 9 from Cuba, 6 each from Puerto Rico, Costa Rica, Brazil, Colombia, and Japan, and 5 from Pakistan and Peru. Today, more than 2879 ASA members live outside the United States in 88 countries (Fig. 1).[1] Since the 1950s, there have been dramatic increases in the number of members from India, Japan, Australia, Germany, and Argentina, but membership has actually declined in a few countries, including Canada, Cuba, and the Dominican Republic. The global distribution of ASA's international members has changed as well. Nearly all members resided in the Americas in 1918 (Fig. 2). By the 1950s, most were still located in the Americas, but several were from the Middle East (including North Africa and Central Asia) and Europe. Since the 1950s, the number and proportion of international members in Southeast Asia (including Oceania) and Europe have increased dramatically, while those in the Middle East and especially Sub-Saharan Africa have remained disappointingly low.

International Activities

Since ASA's inception, its membership has made important contributions to international agronomy. As ASA Historian Wayne Keim (1969) later put it, "In those early years, International Agronomy was of concern to a considerable number of agronomists." Hunnicut (1913) described agronomic concerns in Brazil, Harlan (ASA, 1913) took a six-month furlough from the USDA to evaluate agronomic possibilities near Lake Titicaca in Peru, Bailey (1914) described quality of Mexican wheat, former Assistant Secretary of Agriculture Willet M. Hays served as agricultural adviser to the Argentine government (ASA, 1913), and Bouyoucos (1914) wrote on the agricultural conditions of Germany, France, and England.

In an effort to increase participation, ASA's first meeting took place with the annual meetings of the American Association for the Advancement of Science (AAAS), which is today the world's largest scientific society. ASA continues to take a leadership role in section O of AAAS (Keeney, 1994) and is now officially affiliated with more than 80 international societies. Furthermore, since the development of guidelines in the mid 1960s for the cosponsorship of conferences, seminars, and symposia (Smith, 1980), ASA has cosponsored several international meetings. Examples include cosponsorship of the annual meeting of the American Society of Range Management in 1964, a conference on Mechanization of Field Experiments in 1975 with the American Society of Agricultural Engineers, the International Symposium on Remote Sensing in 1984 with three other scientific societies, an International Conference on Global and National Resource Monitoring and Assessment in 1989 with Crop Science Society of America, an international conference on Advancing Biotechnology in 1990 with USAID and AAAS, a conference on Information Agriculture with the Potash and Phosphate Institute in 1995, a conference on Sustainability of Agricultural Systems in Transition in 1996 with the World Bank, and the scientific congress of the European Society of Agronomy to take place in 2008.

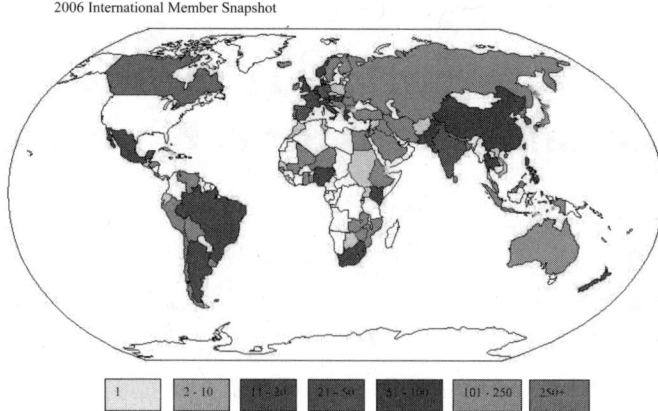

Fig. 1. 2006 international membership distribution.

The series of International Grassland Congresses began in 1920, largely for scientists in northern and central Europe, but the 3rd Congress in Zurich in 1934 was extended to other countries and was truly international and the word was added to the title. The 4th Congress was in Aberystwyth in 1937. After a disruption with the war the 5th Congress was held in 1949 in the Netherlands. Intergovernmental organizations were major supporters. Subsequent Congresses were held about every four years, often outside of Europe, with the 6th in Pennsylvania in 1952, the 14th in Kentucky in 1981, and the 18th in Canada in 1997. The Pennsylvania Congress provided new and expanded opportunities for involvement of ASA members. P. V. Cardon was elected President of the Congress, and W.M. Myers was elected Secretary General. Other leaders at the Congress such as W.A. Minor, G.O. Mott, H.L. Lucas, G.W. Burton and A.A. Hanson were also leaders in ASA, but there was no formal association with ASA until the more recent Congresses. In 1974 R.F Barnes was appointed as the North American representative

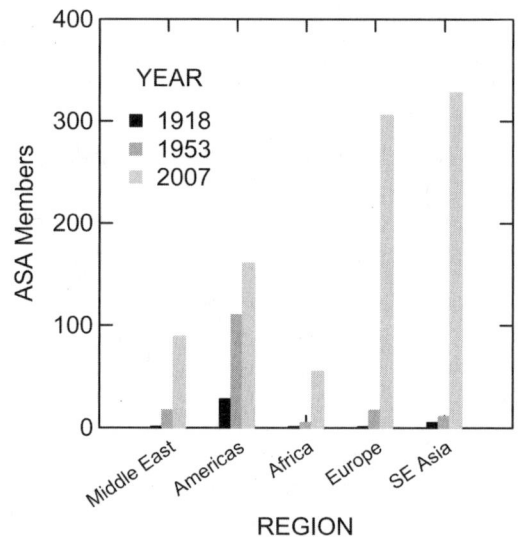

Fig. 2. Global distribution of ASA international members in 1918, 1953, and 2007.

---

[1] Data were compiled from mailing addresses, and may therefore include U.S. citizens living abroad or exclude non-citizens living within the United States.

to the Continuing Committee and served in that role through the 1997 Congress. Several ASA members served on committees for the 1981 Congress. Through his role as Executive Vice-President of ASA, Barnes was instrumental in developing linkages of support from ASA for the Congresses, especially the 1997 Congress in Canada, for which ASA was a Sponsoring Society. At the 2001 Congress in Brazil, V.G. Allen was elected Chair of the Continuing Committee, a role she held through the 2005 Congress in Ireland. At the Ireland Congress, the Continuing Committee voted to place the development and management of a permanent website with ASA. It was further requested that ASA would be the repository for International Grassland Congress funds between congresses. Allen has been active in terminology and in cultivating relationships to develop the important joint meeting with the International Rangeland Congress to be held in China in 2008. ASA is a Sponsoring Society for the joint meeting of the two Congresses.

Publications

In their review of the first 50 years of ASA, Laude et al. (1962) wrote "It might well be said that the publications of a scientific society are the society, for certainly they are the chief tangible evidence of its existence and its vigor and vitality." If such is the case, then international authors have made impressive contributions to the vigor and vitality of ASA. International members currently make up about 15% of ASA's roster, but international authors contributed annually 23 to 31% of the papers published in *Agronomy Journal* (AJ) between 2000 and 2005 (ASA, 2006), and 31 to 40% of submitted manuscripts between 2003 and 2006 (Pearson, 2007). *Agronomy Journal* quite appropriately describes itself on the cover page as "An International Journal of Agriculture and Natural Resource Sciences."

Many ASA publications (see the accompanying CD for a complete list of ASA publications) have addressed fundamental concepts of agronomy relevant to many countries of the world. The range of these publications reflects a gradual shift in emphases with time and coverage of a larger geographical area of the world. Topics included food and related issues of global hunger, agronomy in a world context, soils of the tropics, technology transfer, and international training. In more recent decades, the emphasis expanded to issues such as human nutrition, organic farming, sustainable agriculture in the tropics, sustainability of systems in transition, rice–wheat systems, climate change and agriculture, food production and its environmental implications, and agricultural ethics.

INTERNATIONAL AGRONOMY AND WORLD FOOD SUPPLY

Concern about our capacity to increase world food production to feed a growing world population was first raised by ASA President F.S. Harris (1920) in response to post-World War I food shortages. Harris was not concerned so much with temporary or local conditions of famine as much as "…the means by which people may be fed when the world becomes much more populous than it now is." Addressing an issue that had been raised in Europe more than a century earlier by Thomas Malthus, David Ricardo, and others (Malthus, 1826; Warsh, 2006), Harris (1920) questioned "Is there a limit to the number of people for whom the earth can supply food, or can the increase go on indefinitely?" He believed that increasing production to solve the world's impending food problem would depend heavily on the ability of agronomists to (i) bring more, possibly marginal lands into production and (ii) raise crop yields. He proposed doing the first by increasing irrigated areas, extending dry farming, draining wetlands, and reclaiming alkali land and the second by increasing soil fertility, devising better tillage methods, and improving crops by breeding.

For several years Harris' (1920) views that "The welfare of mankind is intimately bound up with the world's food supply" and that agronomists were vital to its increase went unmentioned in ASA presidential addresses, perhaps due to more pressing problems such as overproduction, diminishing export markets, increased international competition faced by U.S. farmers, the need for soil conservation, and eventually a second world war. Finally, almost 30 years later, ASA President Firman Bear (1949) stated in his presidential address that the Malthusian problem had recently been revived. Although his remarks were largely restricted to U.S. agriculture, he used FAO statistics to encapsulate an impending disaster:

> If the whole world ate in accordance with our standards, not over one-third of its 2¼ billion people could be fed at present world production levels. In comparison with the average diet of all the other people on earth, this more-than optimum diet of ours is nothing less than extravagant.

Bear's (1949) remarks were made shortly after Harry S. Truman's famous "Four Point" inaugural speech, given in January of that same year. Truman's last point was:

> Fourth, we must embark on a bold new program for making the benefits of our scientific advances and industrial progress available for the improvement and growth of underdeveloped areas.
> More than half the people of the world are living in conditions approaching misery. Their food is inadequate. They are victims of disease. Their economic life is primitive and stagnant. Their poverty is a handicap and a threat both to them and to more prosperous areas.
> For the first time in history, humanity possesses the knowledge and skill to relieve the suffering of these people.

By the mid 1950s, efforts of ASA agronomists to increase international food supply were well underway. In the 1956 ASA presidential address, Johnson (1956) spoke of successful international cooperative projects in Mediterranean countries of Europe and Africa for corn improvement, which were sponsored by FAO and led by M. T. Jenkins of USDA-ARS. A similar effort, also sponsored by FAO, was underway among small grain breeders, and led by J.B. Harrington of the University of Sasketchewan, Canada. Johnson (1956) described "an amazing revelation that scientists can work together without diplomatic restraint and to see in actual practice that scientific cooperation is not hampered by international boundaries."

The next year, Bradfield (1957) spoke directly and at length to the need to increase food production:

> Since World War II the world has shrunk. Countries formerly far away are now only a few hours away by air. As a result we are more conscious of, and more concerned about the plight of agriculture in underdeveloped countries than ever before. We are beginning to see that it is to our advantage as well as to the advantage of these countries to have their agriculture more highly developed and their economies in a sounder position.

Bradfield's (1957) address came during the peak of the Cold War. He appears to be among the first in ASA to tie food to peace, a theme that would grow immensely during the next decade. His words still seem appropriate today: "If the investment in agricultural technology is wisely made, it should continue to yield returns for a long time after the military weapons have found their way to the scrap heap."

To Bradfield (1957), the most important cause of hunger was low yield. He recognized that N was the most commonly limiting yield factor to yield in many varied agricultural settings. He also spoke of underdeveloped countries' low capital reserves, poor transportation facilities, and "antiquated land tenure systems" that "kept agriculture in a straight jacket" and, in extreme cases, resulted in famine and widespread starvation. He noted further that diets in poor countries were protein deficient, which contributed to widespread malnutrition, particularly among infants. Underdeveloped countries had insufficient knowledge of their soil resources and proper management, especially in the tropics, where the largest areas of agricultural potential on the planet remained and soils presented unique management problems. He saw lack of training as the principal reason that agriculture in these countries was underdeveloped, and stressed that the goals of the agronomist working in underdeveloped countries should be to "investigate, demonstrate, educate, and motivate."

TACKLING GLOBAL PROBLEMS THROUGH AGRONOMIC RESEARCH AND DEVELOPMENT

By the 1960s, Harris' (1920) belief that humanity's welfare would be tied to global food supply was widely accepted. Glen Burton (1963) spoke at length on the complex relations between agronomy, the food supply, and peace under the theme of "Food for Peace," which was also the title of ASA's first special publication (Brady et al., 1963). Burton (1963) summed up the relation between feed and peace succinctly: "A hungry man knows no law and experiences no peace. He is ready prey to any idea that promises escape from his misery and is usually incapable of weighing the consequences." Burton's words recall those of two Nobel Peace Laureates: Lord John Boyd Orr, who said "You cannot build peace on empty stomachs" (Fannin, 2003), and Norman Borlaug, who said "As long as there are suffering and lack of food, there will be political uprisings and people killing each other" (Stoeltje, 2003).

Burton (1963) and others recognized that the best solution to hunger was to help societies to improve their agricultural output until they could feed themselves. He described successes in Mexico, where agricultural output during the previous 10 years had gone from insufficiency to near self-sufficiency, and where the per capita diet had increased from a substandard level of 1700 calories in the late 1930s to an adequate 2700 calories in 1958. Burton advocated a "blueprint" that called for agronomists or teams of crop and soil scientists to, in the words of M.A. Carleton (1910), "thoroughly investigate anything and everything concerned with the field crop." Burton (1963) saw the need for an entire spectrum of research and chastised those suggesting that applied and development research should be discontinued in favor of basic research. He realized that the highly successful American "blueprint" for agricultural production would not be transferable without modification and understanding the culture of those whom they would teach.

The 1960s were a time of immense progress in international agronomy and its quest to increase food production in an increasingly populous world, with several members of ASA leading the way. We could not possibly cover all the accomplishments or do justice to the many individuals and agencies who contributed, but we have attempted to provide brief biological sketches of some key luminaries who have been recognized publicly for exceptional service to international agriculture and the quest for global food security. An overview of considerable contributions of U.S. federal agencies and Land Grant Universities during this period was given by Smith (1968).

INTERNATIONAL AGRONOMIC EDUCATION

Training of both U.S. and non-U.S. agronomists in international agronomy was addressed in one of the earliest ASA Special Publications (Cowan and Robertson, 1969), in which Milford (1969) stated: "By consensus, the most valuable learning experience for students with an interest in international agronomic problems is a period of work and study in a developing country." The *Journal of Natural Resources and Life Sciences Education*, published by ASA, is described as an "international journal." While there is a paucity of international articles published in JNRLSE, many contributions touched on vital issues in the interface between educators in the USA and those involved in international agriculture and are of enormous benefit to teachers of international agriculture in U.S. universities.

THE EVOLUTION OF ASA'S INTERNATIONAL AGRONOMY DIVISION (A-6)

D.C. Smith (1980) described the formation in 1962 of a sectional program on "Pan-American Agronomy," and subsequent travel by ASA's executive secretary, Matt Stelly, to Latin American countries. Visits were made to the newly founded Association Latinoamericana de Fitotecnia (ALAF) in Cali, Colombia, with support from both ALAF and the Rockefeller Foundation. Following this successful association, ASA considered the feasibility of a more active effort toward international agronomy (McVickar, 1973). In 1967, ASA President Robert S. Whitney appointed five members to a special committee on International Agriculture. The members were L.V. Crowder, M.S. Drosdoff, K.T. Payne, E.J. Wellhausen and M.B. Russell, who chaired. Their charge was "to study the suitability of an International Program in ASA and prepare specific recommendations."

The committee sent a resolution to key people in government, universities, foundations, academics, and industries concerning food supply and demand. It facilitated provision of printed material to agronomic societies in other countries, established the Norman E. Borlaug Book Fund in 1971, and approved editions of Society books in other languages. Chairman Russell also wrote to all agronomic department heads, asking them to designate two or three individuals concerned with international programs. Thirty were selected to attend a discussion session held at the annual meetings in Washington, DC in 1967. In 1971, the division was granted provisional division status by the ASA board of directors, with L.V. Crowder serving as chair. In 1972 the division's permanent designation, A-6, was granted. As stated in its bylaws, the new division's objectives were:

- To encourage excellence in agronomic education and scientific investigation
- To develop programs with agronomic societies of other nations
- To recognize outstanding contributions to the field of international agronomy
- To encourage the exchange of scientific information and technology between scientists of the USA and foreign countries
- To develop the desire for and the respect of agronomic science and for quality in science

GLOBAL/INTERNATIONAL RECOGNITION AND AWARDS

ASA's International Service in Agronomy Award was instituted in 1968 to recognize individuals who had made significant contributions in the area of international agronomy and crop science (see the list of award winners on the accompanying CD). Its first recipient was Nobel Peace Laureate Norman Borlaug. Several international members received the highest recognition of ASA, by being elected Fellow (see the complete list of Fellows on the CD). Finally, a very select group of ASA scientists have received the World Food Prize, which is perhaps the pinnacle of recognition for a lifetime dedicated to improving world food security.

LATER YEARS: CHANGING PERCEPTIONS AND PRIORITIES

Since the 1970s, international agronomy and the need to ensure global food security in the face of rising population and limited land resources has remained a prominent theme of numerous conferences, publications, and presidential addresses. Cowan (1973), for example, traced the path of wheat germplasm from old Kansas varieties imported from eastern Europe (Malin, 1944) to Japanese scientists in Northern Honshu, to Orville Vogel in Washington State University, and on to Norman Borlaug in Mexico, where enormous impact was achieved. He also recognized similar success with rice through research efforts of Robert F. Chandler and others at IRRI. Cowan (1973) captured the success of what was becoming known as the "Green Revolution":

> …these new cereal varieties and farming techniques are paying off very effectively in reducing hunger and malnutrition. They have paid off too in economic gains which bolster the confidence and ability of nations to handle otherwise seemingly insoluble problems. The large harvest has strengthened the people's faith in modern technology and its potential for improving the well-being of their nations. For the first time in centuries, some farmers in India and Pakistan are producing more than is needed for their subsistence and hence are exploring the possibility of mechanization which adds another dimension to their economy.

McCloud (1975) was concerned about global carrying capacity, which he estimated to be approximately 5.8 billion people. He viewed insufficient dietary protein as the most serious food shortage and promoted the urgent need for a "Green Revolution" in the food legumes. Patterson (1977) saw increasing education, research, and development as keys to increasing production, and suggested that a much higher investment would be required to reach realistic food goals by the year 2000 than had been required to reach current food production in 1976. He noted that "Our investment in research has been declining recently rather than increasing." Food-deficient nations, in his view, should (a) develop the technology to produce their own food or (b) raise their industrial economy to afford to buy it. Echoing Glen Burton's (1963) views on self-sufficiency, he stated

> Some food may be available to share on a humanitarian basis by the international community, but it is not assured. The best investment by the international community in these countries is in agricultural research and technology aimed at increasing the food supplies within the countries.

Pendleton (1979) saw a need for annual production increases of 3 to 4% in many developing countries for several decades and listed three central issues to avoiding great chaos and starvation: (i) stockpiling huge food grain reserves, (ii) the introduction of appropriate agricultural technology to allow greater local food production, and (iii) the control of population growth. Pendleton saw an ever-increasing role for agronomists in the first two and called for increased governmental support of research and development.

Three years later, Olsen (1982) spoke about food security and world food distribution. He observed that world food production continued to increase but not fast enough to keep pace with population. At the time of his address, hunger was a daily fact of life for nearly one-quarter of the 4 billion people. Another half billion were seriously malnourished. Food aid from the United States could not make up the deficit, since at that time only about 5% of the world's food supply was transported between countries, mostly by the United States. Olsen's (1982) address reflected an evolving view that hunger was not caused merely by poor agricultural production, but had other economic, political, social, and cultural dimensions. Economic and political empowerment were increasingly seen as more important to reversing hunger, as was the view of Amartya Sen (1984, 2002), who was awarded the Nobel Prize for Economics in 1998. Olsen (1982) spoke of geographic hunger and malnutrition despite overall sufficient world food production. Olsen (1982) wondered whether rates of increase could be sustained, stating that many scientists thought we had nearly reached the biological yield potential of many crops. It was during this period that concern about "yield plateaus" was growing.

Larson (1986) noted that, despite pockets of famine in Africa and chronic malnutrition in other developing countries, total world food production had increased 23% in less than a decade. Citing population projections and trends in the number of persons per hectare of cultivated land, Larson (1986) identified three reasons why soil resources may not be provide an adequate diet worldwide through the expansion of cultivated areas. First, because of social and economic forces, only a 25% increase in cultivated land was feasible. Second, soil degradation in some areas was taking place at an alarming rate because of population pressures. Third, uncultivated lands were far from heavy population centers. He believed that bringing new lands into cultivation would not be enough. Increasing productivity of presently cultivated land was, in his view, an economically and more environmentally sound alternative.

Speaking at the beginning of the biotechnology revolution, Frey (1985) expressed optimism that technology could keep pace with the food demands of a growing population, so long as it was applied in an interdisciplinary context:

> It is my thesis that biotechnology will force the researchers in environmental and genetical disciplines of agronomy to work on cooperative teams. And from this team effort will come an enormity of crop production that may dwarf the accomplishments of the "Green Revolution."

Recalling that it had taken 50 years to understand and completely utilize the Al tolerance gene in wheat, Frey (1985) stated:

> If soils researchers and plant breeders had cooperated on this project as a team, the problem probably could have been solved in a much shorter time—and to greater benefit to humankind. …Yes, the unifying thread will again be DNA. No, the new biotechnology will not make researchers on soils and climate and plant breeders into agronomists, but if the scientists in these disciplines really wish to participate in feeding hungry people, they must join together into agronomy research teams. The future for agronomy is not only bright, but it has no foreseeable bounds.

Moss (1987) was concerned that there were many in Africa and other parts of the world that had not yet shared in the bounty. He recognized the importance of cultural barriers and use of appropriate technology:

> We have come to know that they can't use our large tractors, our huge plows, or in most cases, even our cherished cultivars and fertilizers and management schemes. We found also that our minds are so set by our cultural background and the environments of our countries and soils that often we do not do well at designing the appropriate experiment in third world settings. What they do need from us is our science. We must provide that motivation and technical knowledge to the bright young minds—in those nations where agronomy is not yet an established and time-proven profession.

Nielson (1992) had a somewhat different view of agronomy as a science and its application to other countries, but he recognized the importance of culture:

> We tend to teach the art and practice of agronomy at the local or state level rather than teaching first rate basic science, and examples of its application to technology. And the examples should be expanded to regional, national and global perspectives. When mathematics, physics, and chemistry are taught, the principles and their application are made worldwide....How do we teach agronomy, crops, and soils? Too often, under empirical local conditions that makes their application regionally or globally, most difficult. First, agronomy really depends on the people.

Nielson (1992) described

> …Above all, a management system must be developed that is geared more to sustainable development, especially in third-world countries, than to the present uncoordinated interests and actions of our so-called developed countries. In the American Society of Agronomy, with 20% of our membership from other

countries and with the increased role and vision of the industrial and practicing agronomist, this ... is the key, most important global agronomic opportunity. We must act and be more creative at both national and international levels.

Nielson was perhaps the first ASA president to speak of global, sustainable development. Nielson's (1992) remarks coincided roughly with the acceptance of sustainability as an overarching theme in agriculture and other disciplines concerned with global development. "Sustainability" has been defined differently among various groups and disciplines, but within the context of agricultural systems, most published definitions (Payne et al., 2001) include components of favorable economics, conservation of resources, preservation of ecology, and promotion of social justice.

Acceptance of "sustainability" as an overarching theme reflected a growing consensus that yield increase alone would not be sufficient to address global development challenges, and that other social and environmental considerations would have to be addressed as part of "sustainable" solutions. Even in the 1970s, doubts about universal success of the Green Revolution were being raised. For example, concerns about social equity were expressed by ASA President Blaser (1971):

> This outstanding biological achievement in plant breeding and soil management by members of ASA is now broadening the gap between the rich and the poor. The small peasant farmer cannot compete with large operators because of deficiencies in capitalization and management skills. Thus, additional migration of farm people to already overcrowded cities with serious unemployment is causing violent political and economic upheavals... That the green revolution is causing almost as many problems as it is solving does not minimize the importance of *scientific* achievement.

Others raised concerns about the Green Revolution's impact on the environment (FAO, 1996). Early on, Patterson (1977) raised issues of coping with energy and environmental problems, and the need to consider "food supply impact" as well as "environmental quality impact."

THE FUTURE OF INTERNATIONAL AGRICULTURE AND ASA

Keeney (1994) predicted that ASA's international ties would continue to be built through memorandums of understanding with several countries, and members in all facets of international agriculture. He predicted that ASA's expertise would be increasingly called on to meet the demands for food and fiber and improve the environment worldwide, and saw these international bridges as critically important to the future of the Society. He viewed the new world order for agronomists as one that involved being more responsive to user and societal needs, including food safety, natural resource protection and enhancement, and food security nationally and internationally. Similarly, Foss (1996) saw a growing international role for agronomists with international accords, and in maintaining U.S. agriculture's presence in world markets. He observed that, economically, agriculturally, and environmentally, we are in an increasingly global setting, and that our teaching and research programs should place more emphasis on the global aspects of agriculture. The challenges for international agriculture, then, as well as the number of positive roles that ASA agronomists can potentially play, would seem to be as great as ever.

There are, however, some troubling trends. Despite the warnings of many prominent voices on the importance of sustained commitment to international agricultural research to address issues of food security and sustainable development, public investment in international agricultural research and development has been stagnant or declining. For example, USAID's agricultural funding declined from a high in 1985 of about $1.2 billion to a low in 1997 of only $145 million (IRG, 2005), and has since trended up only slightly. Among CGIAR centers, unrestricted funding has been constant or perhaps only slightly increasing over the past 5 yr. Restricted funding has greatly increased, and became the source of much competition among CGIAR Centers. However, 82% of restricted funding, or nearly $200 million, was designated as project funding, which is the most restrictive in terms of its use, and a considerable percentage does not support CGIAR System priorities (Task Force On Funding System Priorities, 2005). But System priorities were set as part of a long-term strategy to address the very problems for which centers were created–often they are among the most intractable in the agronomic sciences. It is difficult to envisage reversing the trends of poverty, malnutrition, hunger, and land degradation in the remaining underdeveloped areas of the world, particularly in Sub-Saharan Africa, with such little public commitment of funds, the bulk of which consists of short-term projects not always directly related to solving long-term objectives. The trends of underfunding international agricultural research and development ignore Bradfield's (1957) counsel that investment in agricultural technology would yield returns much longer than investment in weaponry, Burton's (1963) admonition against reducing funding in applied and development research in favor of basic research, and warnings of Patterson (1977) and Pendleton (1979) on the dangers of under funding agricultural research. Even the promise of modern biotechnology is unlikely to be realized, as Frey (1985) point out more than 20 years ago, if it is not applied within the context of agronomic and ecological systems.

There are worrisome trends within ASA, as well. The number of international members has declined in both absolute and percentage figures since 1997, and especially since 2001 (Fig. 3). (The slight increase 2005 coincided with a reduction in membership costs to underdeveloped countries). One can only speculate as to the causes. It may have to do with the separation of the tri-societies, the September 2001 attacks and subsequent visa restrictions, or the rise of other international agronomic societies, such as the European Society of Agronomy. But whatever the reason, international members appear to see less and less value in joining and remaining part of ASA. Some thought should be given to reconciling the dual trends of increasing international contributions to Agronomy Journal, on the one hand, and decreasing international membership in ASA, on the other. As Fig. 1 shows, international membership is especially lagging in the Middle East and Africa. Based on the "bump"

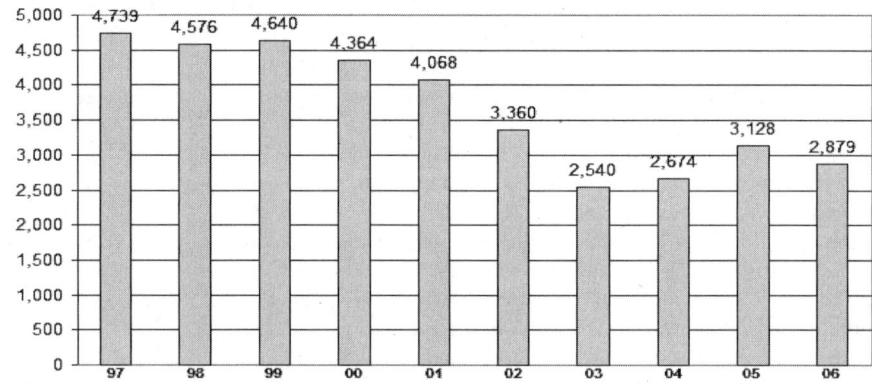

Fig. 3. *Recent trends in international membership for the combined Societies.*

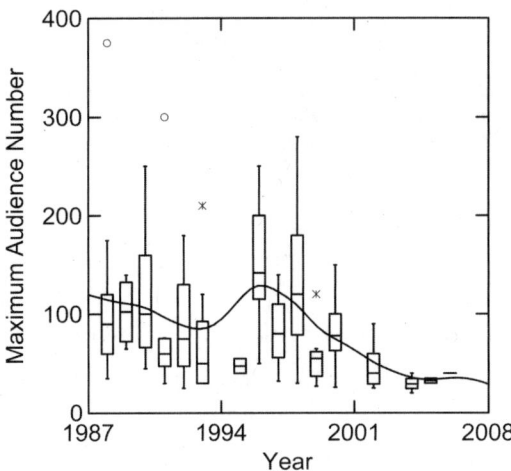

Fig. 4. Box and whisker plot of maximum attendance at A-6 sessions at ASA annual meetings, as reported by session chairs. The data indicate a drop in both the number of sessions and the number of sessuib attendees, especially since 1998. No data were reported for 1994, 2001, and 2003.

in international membership observed in 2005 (Fig. 3), and the small current number of ASA members from these two regions, a drastic decrease in membership fees for citizens of these countries might encourage increased membership without serious financial implications for ASA.

Trends for division A-6 are even more troublesome, as shown by the precipitous drop in the number of sessions at the annual meetings, and the mean number of attendees at these sessions (Fig. 4). Despite its proud past, in which A-6 chairs regularly corresponded with congress, cabinet secretaries, and leaders of other countries' agronomic societies, this division has become a shadow of its former self. This too may be in part due to separation of the tri-societies into ASA, CSSA, and SSSA, for certainly international papers are given within other divisions. Or it may in part be due to the decrease in USAID-funded graduate students attending the meetings. But whatever the cause, the level of enthusiastic, well coordinated effort that once took place within this division to address issues of food security and sustainable development has greatly diminished, even while the scope and magnitude of the challenges it was designed to address remain as daunting as ever. Revitalization of the international division is desperately needed, and ways of effecting this deserve some thoughtful consideration. Possibilities include Web-participation by international members in A6 or other ASA sessions, the creation of electronic newsletters, renewal of formal links and outreach to national programs, and the extension of free membership and online journal prescription to select scientists, such as Director Generals, of national agricultural research entities in developing countries.

SUMMARY AND CONCLUSIONS

The history of ASA's international involvement is a long and proud one that could not possibly be adequately covered given space limitations. Since its earliest days, ASA has both been influenced by and contributed to agronomy in other countries. Initially, almost all international members were from the Americas. Today, there are more than 2800 members living in 88 countries, mostly in Southeast Asia, Europe, and the Americas. International members currently make up about 15% of ASA's roster, but each year they contribute ~20–30% of the papers published in Agronomy Journal, and ~30–40% of submitted manuscripts.

ASA has been officially affiliated with other international societies since its very first meeting, and today is affiliated with more than 80 societies that are dedicated to agricultural research. Additionally, it has cosponsored several international meetings, and contributed a great many publications that have had direct or indirect scientific relevance to agronomy in many if not most countries of the world.

Much of the motivation to participate in international agronomy had to do with world food production to feed a growing population. This was first mentioned in 1920 by ASA president F.S. Harris, but did not really take hold until the 1940s, and especially after Harry Truman's 1949 "Point Four" inaugural speech. By the mid 1950s the seeds of the "Green Revolution" were firmly planted and the idea that food supply and peace were connected had been widely accepted. The 1960s were a time of immense progress in international agronomy and its quest to increase food production in an increasingly populous world, with several members of ASA leading the way.

It was during the late 1960s that ASA began considering the creation of a division devoted to international agronomy. This was realized in 1972. Several awards and honors, including chairs, board reps, fellows, and the prestigious International Service in Agronomy Award, were instituted around this time. Since the 1970s, international agronomy and the need to assure global food security in the face of rising population and limited land resources have remained prominent themes in ASA, and have been the topic of numerous conferences, publications, and presidential addresses. However, there has also been an evolving view that hunger was not caused merely by poor agricultural production, but had other economic, political, social, and cultural dimensions. A belief in the role of technology in meeting food demand remains, especially for new ones such as molecular biology, but these must be applied in a "systems" context. Since the 1990s, concepts of "sustainable development" have emerged and matured. Despite the many successes achieved in international agronomy, many parts of the world have yet to see a "Green Revolution," especially Sub-Saharan Africa.

There remain many challenges and opportunities to address in the area of international agronomy and sustainable development, including being responsive to societal demands in the areas of food safety, natural resource protection and enhancement, and food security. These will require active involvement of ASA members. There are, however, some troubling signs with regard to international membership and especially the level of activity of A-6, the international division. We have given some suggestions as to how these trends might be reversed, but above all we need the continued influx of bright young and idealistic minds to ensure that the world's population is fed and clothed without loss of the resource base or damage to our global environment. We hope that this brief expose will help to encourage future generations to meet this noble challenge.

Acknowledgments

The authors would like to thank ASA HQ staff, and especially Sara Uttech, for all their help with gathering membership data, providing archived material, and general support. We would also like to thank the many contributors. Finally, we thank Lowell Moser and Ed Runge for many helpful comments on earlier manuscript drafts.

REFERENCES

ASA. 1911. List of present members, with addresses. Proc. Am. Soc. Agron. 2:19–23.

ASA. 1913. Agronomic affairs. J. Am. Soc. Agron. 4:251.

ASA. 1953. List of present members with addresses. Agron. J. 45:341–352.

ASA. 2006. Reports of ASA divisions, branches, and committees, 2005. Agron. J. 98:840–848.

Bailey, C.H. 1914. Composition and quality of Mexican wheats and wheat flours. J. Am. Soc. Agron. 5:57–64.

Bear, F.E. 1949. Food for thought about food. Agron. J. 41:497–507.

Blaser, R.E. 1971. Presidential address—Broadening horizons. Agron. J. 63:1–3.

Bouyoucos, G. 1914. Some observations on the agricultural conditions in Germany, France and England. J. Am. Soc. Agron. 6:139–159.

Bradfield, R. 1957. The agronomist's accomplishments and opportunities for future contributions in the International field. Agron. J. 4:621–625.

Brady, N.C., A.A. Hanson, and M. Stelly (ed.) 1963. Food for peace. ASA Spec. Publ. 1. ASA, Madison, WI.

Burton, G.W. 1963. ASA presidential address—The Agronomist and Food for Peace. Agron. J. 55:1–3.

Burton, G.W. 1963. The agronomist and food for peace. p. 1–5 In N.C. Brady et al. (ed.) Food for peace. ASA Spec. Publ. 1. American Society of Agronomy, Madison, WI.

Carleton, M.A. 1910. Development and proper status of agronomy. Presidential address at the Washington meeting, 1908. Proc. Am. Soc. Agron. 1:17–23.

Cowan, J.R. 1973. Presidential Address. The seed. Agron. J. 65:1–4.

Cowan, J.R., and L.S. Robertson (ed.). 1969. International agronomy training and education. ASA Special Publ. 15. ASA, CSSA, and SSSA, Madison, WI.

Fannin, B. 2003. Borlaug discusses future world food production at seminar. AgNews and Public Fairs, October 2003 issue. Texas Agriculture Exp. Stn., College Station, TX.

FAO. 1996. Towards a new Green Revolution. Proceedings of World Food Summit. Food for all. 13–17 Nov. 1996. FAO, Rome.

Foss, J.E. 1996. Presidential address. Professional contributions. Agron. J. 88:119–121.

Frey. 1985. Presidential address. The unifying force in agronomy—Biotechnology. Agron. J. 77:187–189.

Harris, F.S. 1920. The agronomist's part in the world's food supply. J. Am. Soc. Agron. 12:217–225.

Hunnicut, B.H. 1913. Some Brazilian problems in agronomy. J. Am. Soc. Agron. 4:34–38.

IRG. 2005. Agriculture and natural resources management research priorities desktop review. EPIQ II IQC. International Resources Group, Washington, DC.

Johnson, I.L. 1956. Teamwork in agronomy. Agron. J. 48:535–537.

Keeney, D.R. 1994. Presidential address. Building bridges: The American Society of Agronomy and the world. Agron. J. 86:219–220.

Keim, W.F. 1969. The developing years of the American Society of Agronomy. Unpublished ASA Historian report (File A232). ASA, Madison, WI.

Larson, W.E. 1986. Presidential address. The adequacy of world soil resources. Agron. J. 78:221–225.

Laude, H.H., M.F. Miller, J.D. Luckett, G.G. Pohlman, D.S. Metcalfe, W.H. Pierre, and E. Truog. 1962. History of the American Society of Agronomy. First fifty years—1907–1957. Agron. J. 54:57–69.

Malin, J.C. 1944. Winter wheat in the Golden Belt of Kansas. A study in adaptation to subhumid geographical environment. University of Kansas Press, Lawrence.

Malthus, T.R. 1826. An essay on the principle of population: A view of its past and present effects on human happiness; with an inquiry into our prospects respecting the future removal or mitigation of the evils which it occasions. 6th ed. John Murray, London.

McCloud, D.E. 1975. Presidential address: Man and his food. Agron. J. 67:1–3.

McVickar, M.H. 1973. Report—Division A-6 (International Agronomy). Unpublished report. ASA, Madison, WI.

Milford, M.H. 1969. The American agronomy student discovers the world. p. 1–10 In D.C. Smith (ed.) Food for billions. ASA Special Publ. 11. ASA, Madison, WI.

Moss, D.N. 1987. Presidential address. Agronomy in a global economy Agron. J. 79:185–187.

Nielson, D.R. 1992. Presidential address. Global agronomic opportunities. Agron. J. 84:131–132.

Olsen, S.R. 1982. Presidential address. Removing barriers to crop productivity. Agron. J. 74:1–4.

Patterson, F.L. 1977. Presidential address. Agronomists and food—Contributions and challenges for the cereals. Agron. J. 69:1–4.

Payne, W.A., D. Keeney, and S. Rao. 2001. Sustainability of agricultural systems in transition. ASA Spec. Publ. 64. ASA, CSSA, and SSSA, Madison WI.

Pearson, C.H. 2007. Letter from the editor. Agron. J. 99:320–321.

Pendleton, J.W. 1979. Presidential address. The road ahead. Agron. J. 79:1–6.

Sen, A. 1984. Poverty and famines: An essay on entitlement and deprivation. Oxford Univ. Press, Oxford.

Sen, A. 2002. Why half the planet is hungry. Observer of London. June 16, 2002.

Smith, D.C. 1980. Development of the American Society of Agronomy, 1958–1977. Agron. J. 72:227–240.

Smith, D.C. (ed.) 1968. Food for billions. ASA Spec. Publ. 11. ASA, Madison, WI.

Stoeltje, M.F. 2003. The man who fed millions. Available at http://soilcrop.tamu.edu/teaching/borlaug.html (verified 28 June 2007).

Task Force On Funding System Priorities. 2005. Task Force on Funding System Priorities. Coordinated action by CGIAR members. Final report for the annual general meeting 2005. Available online at CGIAR.org (verified 28 June 2007).

Warsh, D. 2006. Knowledge and the wealth of nations. W.W. Norton & Co., New York.